For most of us, design is invisible.
Until it fails.

Transmission towers crushed and power lines downed by freezing rain in Boucherville, Québec, January 1998.

In fact, the secret ambition of design is to become invisible, to be taken up into the culture, absorbed into the background. The highest order of success in design is to achieve ubiquity, to become banal.

Burned-out control room of Reactor 4, Chernobyl, Russia, June 2001.

The automobile, the freeway, the air-plane, the cell phone, the air conditioner, the high-rise – all invented and developed first in the West, but fully adopted and embraced the world over – have achieved design nirvana. They are no longer considered unnatural. They are boring, even tedious.

Highway damaged by the Northridge earthquake, California, January 1994.

Most of the time, we live our lives within these invisible systems, blissfully unaware of the artificial life, the intensely designed infrastructures that support them.

Hog facility flooded by Hurricane Floyd, North Carolina, September 1999.

Accidents, disasters, crises. When systems fail we become temporarily conscious of the extraordinary force and power of design, and the effects that it generates. Every accident provides a brief moment of awareness of real life, what is actually happening, and our dependence on the underlying systems of design.

Jet fuel tank fires caused by Supertyphoon Pongsona, Apra Harbor, Guam, December, 2002.

Every plane crash is a rupture, a shock to the system, precisely because our experience of flight is so carefully designed away from the reality of the event. As we sip champagne, read the morning paper, and settle in before takeoff, we choose not to experience the torque, the thrust, the speed, the altitude, the temperature, the thousands of pounds of explosive jet fuels cradled beneath us, the infinite complexity of onboard systems, and the very real risks and dangers of takeoff and landing.

Reconstructed wreckage of Boeing 747, TWA flight 800, Long Island, New York, November 1996.

Massive Change is an ambitious project that humbly attempts to chart the bewildering complexity of our increasingly interconnected (and designed) world. We have done our best to open it up by breaking it down, and putting as many fascinating fragments as we could find back together again, between the covers of this book. We hope to make evident the design decisions that go on and are made manifest across disciplines. Massive Change is not about the world of design; it's about the design of the world.

Cars crushed by the Northridge earthquake, California, January 1994.

Institute without
Boundaries 2003
Vannesa Ahuactzin
Mark Beever
Lorraine Gauthier
Jennifer Leonard
Tyler Millard
Alejandro Quinto
Jeremy Everett
Institute without
Boundaries 2004
Leilah Ambrose
Doug Chapman
Gisele Gass
Tobias Lau
Jill Murray
Jason Severs
Ilene Solomon
Britt Welter-Nolan

MASSIVE

Program Director
Greg Van Alstyne
Coordinator
Helen Papagiannis

Bruce Mau
with
Jennifer Leonard
and the
Institute without
Boundaries

Φ

Phaidon Press Limited
Regent's Wharf
All Saints Street
London N1 9PA

Phaidon Press Inc.
180 Varick Street
New York, NY 10014

www.phaidon.com

First published 2004
© 2004 Phaidon Press Limited

ISBN 0 7148 4401 2

A CIP catalogue record for this book
is available from the British Library.

Designed by Bruce Mau Design, Inc.
Printed in China

Bruce Mau Design
Since its inception in Toronto in 1985, Bruce Mau Design
has gained international recognition for innovation in
a wide range of areas, including identity and branding,
research and conceptual programming, print design and
production, environmental design, exhibition design,
and product development for Maharam and Umbra.
Major attention focused on the studio in 1995 with the
release of *S,M,L,XL*, a 1,300-page compendium of
projects and texts designed and conceived by Bruce Mau
Design with architect Rem Koolhaas and his Office for
Metropolitan Architecture. *Life Style*, Mau's monograph
on design culture and the work of the studio, was published
by Phaidon Press in 2000. Bruce Mau Design has worked
with a roster of prestigious clients, including Frank
Gehry, the Getty Research Institute, Gagosian Gallery,
Indigo books in Canada, and The Andy Warhol Museum.

The Institute without Boundaries
Our aim is to produce a new breed of designer, one
who is, in the words of R. Buckminster Fuller, a "synthesis
of artist, inventor, mechanic, objective economist, and
evolutionary strategist." Each year we will select from
an international pool a small group of postgraduate
students who will spend one year inside the Bruce Mau
Design studio, working as a team to research, design,
and realize a public intellectual project. Students are
drawn from a diverse range of backgrounds to include
artists, scientists, journalists, architects, filmmakers,
and entrepreneurs – all of whom bring experience to the
realm of global design.

Now that we can do anything, what will we do?

The twentieth century will be chiefly remembered by future generations not as an era of political conflicts or technical inventions, but as an age in which human society dared to think of the welfare of the whole human race as a practical objective.
                    – Arnold J. Toynbee, English historian (1889–1975)

In his Nobel Prize acceptance speech on December 11, 1957, former Prime Minister of Canada Lester B. Pearson quoted historian Arnold Toynbee, well known for his monumental *A Study of History*. The main thesis of Toynbee's work is that the well being of a civilization depends on its ability to respond creatively to challenges, human and environmental. He was optimistic about the twentieth century. He believed that the cycle of rise and decline was not inevitable, that the future is not determined by the past, and that a civilization could choose and act wisely in the face of recurring hardships. His prediction posed a challenge – and an opportunity – during the post-WWII era; it was significant enough for Pearson to reference it in the context of international peacekeeping during the Cold War era; and it continues to challenge us today, into the twenty-first century.

Our world now faces profound challenges, many brought on by innovation itself. Although optimism runs counter to the mood of the times, there are extraordinary new forces aligning around these great challenges, around the world. If you put together all that's going on at the edges of culture and technology, you get a wildly unexpected view of the future. Massive Change charts this terrain.

We will explore design economies.
Not since the age of invention have so many new products, processes, and services become available to the public. What we see over the last hundred and fifty years, and in a dramatically accelerated pace over the last fifty, is that design is changing its place in the order of things. Design is evolving from its position of relative insignificance within business (and the larger envelope of nature), to become the biggest project of all. Even life itself has fallen (or is falling) to the power and possibility of design. Empowered as such, we have a responsibility to address the new set of questions that go along with that power. At the same time, we acknowledge the hubris and inherent paradox of the new position we find ourselves in: We are designing nature and we are subject to her laws and powers. This new condition demands that design discourse not be limited to boardrooms or kept inside tidy disciplines. As a first step to achieving this, we abandoned the classical design disciplines in our research and, instead, began to explore systems of exchange, or design "economies." Instead of looking at product design, we looked at the economies of movement. Instead of isolating graphic design, we considered the economies of information, and so on. The patterns that emerged reveal complexity, integrated thinking across disciplines, and unprecedented interconnectivity.

We will tap into the global commons.
Massive Change is about the power and promise of design.
Design success equals global success. What makes this
possible is the radical change in scale in the capacity of
design to meet human needs the world over. Extraordinary
projects are underway that are changing our world for
the greater good. Many of the people we include in Massive
Change do not consider themselves designers. But, if you
listen carefully, they (and others like them) use the word
design to describe their work; they speak about designing
systems, designing organizations, designing organisms,
designing programs. We must applaud and participate in
the efforts of these thought leaders (and doers) — or risk
losing them. There is an incredible story to be told about
human ingenuity!
The first step to its unfolding is to reject the binary notion
of client/designer. The next step is to look to what is
going on, right now. The old-fashioned notion of an
individual with a dream of perfection is being replaced
by distributed problem solving and team-based multidisciplinary practice. The reality for advanced design
today is dominated by three ideas: distributed, plural,
collaborative. It is no longer about one designer, one
client, one solution, one place. Problems are taken up
everywhere, solutions are developed and tested and
contributed to the global commons, and those ideas are
tested against other solutions. The effect of this is to
imagine a future for design that is both more modest and
more ambitious. More modest in the sense that we take
our place in what our studio's chief scientist Bill Buxton
calls the renaissance team, a group that collectively
develops the capacity to deal with the demands of the

given project. More ambitious in that we take our place in society, willing to implicate ourselves in the consequences of our imagination.

We will distribute capacity.
One thing is certain: We don't need a thought police. We need discussion. We need thinking. We need critical faculties. We need to embrace the dilemmas and conflicts in design, and take responsibility for the outcomes of our work. When we use the term "we," we don't mean designers as separate from clients, or as some extraordinary class of powerful overseers. We mean "we" as citizens collectively imagining our futures. It is critical that the discussions go beyond the design fields themselves and reach out to the broadest audience, to the people directly affected by the work of designers. The effect of the new conditions is to distribute potential, or capacity, worldwide and allow contribution by anyone, anywhere. The future of global design is fundamentally collaborative. In this condition there is no room for censorship.

We will embrace paradox.
Massive Change calls for greater public discourse and personal responsibility for designers and their projects; at the same time, it is thrilled by the open-source effect of the cultural project of design. The moment we came upon Toynbee's quote in Pearson's lecture, we knew we had our project because it included the phrase "practical objective"; it shifted the objective of the welfare of the human race from a utopian ambition – one that is by definition out of reach and will remain in the realm of art – to a design project, a practical objective.

There is a proposal integrated into Massive Change for a
right-angle shift in the axis of discourse defined by right
and left, socialism and capitalism. The new axis is defined
by advanced and retrograde, forward and reverse. Plainly,
Massive Change is a project that embraces the potential
of advanced capitalism, advanced socialism, and advanced
globalization. In that sense, Massive Change is obviously
ambitiously positive, and might be misunderstood as
utopian at first glance. But it is not futuristic. It is about
what is already happening.

We will reshape our future.
Between 1965 and 1975, R. Buckminster Fuller conducted
five two-year "advanced design science" intensives and
pulled together the results of the participants' research and
analysis into several volumes of what he called the World
Resources Inventory. In Phase I of the work, "Inventory of
World Resources Human Trends and Needs," he wrote,
"There are very few men today who are disciplined to
comprehend the totally integrating significance of the 99
percent invisible activity which is coalescing to reshape our
future. There are approximately no warnings being given to
society regarding the great changes ahead. There is only
the ominous general apprehension that man may be about
to annihilate himself. To the few who are disciplined to
deal with the invisibly integrating trends it is increasingly
readable in the trends that man is about to become almost
100 percent successful as an occupant of the universe."
This book is dedicated to all those with the discipline to
comprehend the total integrating significance of the 99
percent invisible activity which is coalescing to reshape
our future. This is the beginning.

Acknowledgments

Massive Change is the product of the efforts of many individuals and organizations. The project is conceived as a network that grows exponentially every day. For that reason, it's inevitable that someone or many will be neglected. For any of these omissions, I take the sole responsibility.

In May 2002, Bruce Grenville, Senior Curator of the Vancouver Art Gallery, asked our studio to develop and produce an exhibition on the future of design culture. It seemed impossible – the scope and possibilities so infinite. At the same time, the studio was approached by George Brown College in Toronto about developing an education program, which eventually became the Institute without Boundaries, an experimental postgraduate educational initiative whereby students spend a full year in our studio, immersed in research-based multidisciplinary activities. We realized that, by making the Vancouver Art Gallery's exhibition the focus of the first two years of the Institute, we might just be able to pull it off. Inevitably, many other exciting manifestations of Massive Change emerged, including this, our second book with Phaidon Press.

The Institute led the research, development, design, and production of Massive Change. Their spirit, dedication, and talent carried the project. The roster for the first year (2003) included Vannesa Ahuactzin, Mark Beever, Lorraine Gauthier, Jennifer Leonard, Tyler Millard, Alejandro Quinto, and, for a short time, Jeremy Everett. And for 2004: Leilah Ambrose, Doug Chapman, Gisele Gass, Tobias Lau, Jill Murray, Jason Severs, Ilene Solomon, and Britt Welter-Nolan. The Institute's program director is Greg Van Alstyne, whose intellectual and artistic direction are evident in all aspects of the project. Helen Papagiannis joined Greg and the Institute and began research on the project in 2002, participated in the content development, and coordinated myriad components of its various expressions.

Jennifer Leonard, one of the students from the inaugural year of the Institute, stepped up to bat and emerged as the book's editor and my principal collaborator on developing the book's content and composition. Without Jennifer's superhuman commitment, patience, intelligence, wit, and sense of humor, there would be no book.

Christopher Bahry, who led and executed beautifully the design development, and Judith McKay, who successfully managed the challenging schedule and production cycle, joined Jennifer and me at just the right time to make the dream of a book a reality. Alejandro Quinto worked on earlier versions of the book and many others contributed in various ways, including Kelsey Blackwell, Barr Gilmore, Mark Beever, Breanne Woods, Peter Blythe, and William Lam.

Jim Shedden acted as the overall producer of Massive Change, working closely with Greg, Helen, and the members of the Institute; the book team; Vannesa Ahuactzin, who led the charge on the exhibition, and Eric Leyland, who provided valuable project management expertise; Quinn Shephard, my resilient, calm, and patient executive assistant; Cathy Jonasson, Kevin Sugden, Amanda Ramos, Pauline Landriault, Angelica Fox, Sara Weinstein Kohn, Sarah Dorkenwald, Beth Mally, Sarah Newkirk, Ruth Silver, Sonny Obispo, Mark Guertin, Mike Bartosik, Natalie Black, Danella Hocevar, and everyone else in our studio, all of whom made significant contributions to the project.

The primary research for the project emerged out of a radio program that Jennifer developed for the University of Toronto's campus and community radio station, CIUT 89.5 FM, as an extension of her activities with the Institute and the studio. We are grateful to the station, especially station manager Brian Birchell, program director Ken Stowar, and show technician Christine Hirtescu, for making this extraordinary opportunity possible. Thank you also to Chris Warren for transcribing over thirty hour-long transcripts! Digital audio files of each of the original interviews have been archived and are available at the project's website (www.massivechange.com).

This book's content would not be as rich as it is were it not for the generosity of the researchers and professionals (and their assistants) who happily donated images, statistics, and time on the phone fact-checking with Jennifer. Many of you are thanked in the Editor's Note; those who have yet to be thanked are: Teddy Cruz, Miriam Gago of the Institute for Liberty and Democracy; Paulo Krauss; Alexander V. Ermishin of Ekip Aviation Concern; Neal Singer of Sandia National Laboratory; Ana Moore of Arizona State University; Laura Lynch of Stanford Law School; Julie Wulf-Knoerzer of Argonne National Laboratory; Fritz Hasler, Richard McPeters, and Harold Pierce of NASA; Jill Rosenblum and Ariane Bradley of The Natural Step; Susan Mitchell and Valerie Combs of Chaordic Commons, Inc.; Jan Crawford; Ilda Thompson of the Langer Lab at MIT; Paul May of the University of Bristol; Valentin Ryzhov of the Institute for High Pressure Physics in Russia; Fritz Vollrath of Oxford University; Andre Geim of the University of Manchester; Harry Kroto of the University of Sussex; Nicole Grobert of the University of Oxford; Andreas Lendlein of mnemoScience; Scott White of the University of Illinois at Urbana-Champaign; George Lisensky of Beloit College; Eames Demetrios; Julia Campbell of the Institute for Creative Technologies; Caleb Waldorf of the Watson Institute for Global Security; Marcus Schubert; Margaret Sanders of William McDonough & Partners; Steven Goldstein of the University of Michigan; Ian Ambler and Jonathan Todd of Ocean Arks International; John Santini of MicroCHIPS, Inc.; Yadong Wang of Georgia Tech and Emory; Richard Jefferson of CAMBIA; Joanne Veto of Ashoka: Innovators for the Public; Kim Ciabattari and Martin Gross of University of California at San Francisco; and Ji Mi Choi and Gordon McCord of the Earth Institute at Columbia University.

We would also like to thank the entire roster of seminar guests from the weekly in-studio series from the first year of the Institute without Boundaries. Each of you informed the project (in order of appearance, in person, or by way of video conference): Andrew Davies, Petra Chevrier, Patsy Craig, Tom Wujec, Albert-László Barabási, Sabine Himmelsbach, Kent Martinussen, Barry Vacker, Rafael Simon, Albert Nerenberg, David D'Andrea, Richard Hunt, William Thorsell, Richard Garner,

Heather Reisman, Jennifer Elisseeff, Kim Sawchuk, Maggie Orth, Jeff Kipnis, Avigdor Cahaner, Filiz Klassen, Richard Stallman, Ronald Deibert, Ry Smith, David Bornsten, Dan Sturges, and Sue Zielinski. Several seminar participants deserve special thanks: Tony Scherman, for two of the most extraordinary and inspiring presentations I've ever witnessed in my life; John Todd, an inspiration to the Massive Change project; Bill Buxton, the studio's "chief scientist" and advisor on many matters; Ed Burtynsky, who taught us that we can't change anything without paying witness to it and doing it beautifully; Bart Testa, who gave us valuable editorial advice early in the project; and Mike Smith, who helped provide a larger aesthetic context for our work.

The other extraordinary contribution to the content of the book has come from the hundreds of lenders of images, from professional photographers (Ed Burtynsky, Dee Breger, David Malin) to businesses, archives, and many, many individuals who responded to our open call for images. The response was overwhelming and – although not all of your images made it into the book – each of you deserves special mention for your generosity: Juan Ahuactzin, M. Ali Musa, Roger Allan, Michael Alstad, Daniel Abramson, Ozge Acikkol, Laurence Acland, Victor Angelo, Adam Antoszek-Rallo, Julius Aquino, Yves Arcand, Yomar Augusto, Yitzhak Avigur, Adelle Bailey, Rachel Bajada, Roshan Bangera, Mike Barker, Anne Barliant, Nurit Basin, Lise Beaudry, Ijose Benin, Amanda Bennett, Sam Bietenholz, Sylvia Borda, Josh Bortman, Susan Boyle, Carrie Bradish, Terry Brennan, Benton Brown, Derek Brunen, Bumbum (N.T.BINH), Stephen Butson, Susan Callacot, L K Chan, Sharon Chang, Judy Cheung, Sonia Chow, Sarah Chu, Aaron Clark, Constance, Jim Cooke, Sandra Coppin, Manuel Córdova, Colleen Corradi, Chris Culver, Sarawut Chutiwongpeti, Jennifer Wöhrle, Jackie Cytrynbaum, Steven Dagg, Maciej Dakowicz, Michael D'Amico, Andrew Danson Danushevsky, Bruce Danziger, John Darwell, Nancy Davison, Michael De Feo, Adri A. C. de Fluiter, Serge de Gheldere, Edison del Canto, Bianca Dias

DeMartino, John Di Stefano, Susan Dobson, Frederico Duarte, Niki Dun, Marcin Dzieszko, Jeremy Edmunds, Meighan Ellis, Fumino Enokido, Fancy Camera, Hiba Farhat, Catherine Farquharson, Bridget Farr, Jussara Figueredo, Adrian Fish, Maxe Fisher, Jan Flook, Joe Fotheringham, Nelson French, Fume Design, Helle Gade Jensen, Joy Garnett, Sean George, Murat Germen, Ed Giardina, Pablo Gimenez Zapiola, Jose Antonio Gonzalez, Steeve Gosselin, Carole Guevin, André Guyot, Louise Haddow, Julian Haladyn, Patrice Hanicotte, Andrea Hamann, Hariri Pontarini Architects, Arriz Hassam, Isabelle Hayeur, Paul Harrison, Margaret Heffernan, Chuck Hemard, Virginia Hilyard, Hitchco, Debbie Ho, Chan Hoi Yoen, Lorens Holm, Ihor Holubizky, Ricardo Hubbs, Rich Hudnut Jr., Matthew Hufft, Gudrun Hughes, Eugene Ilchenko, Luis Jacob, David Jacobson, Erfanian Javan, Margarita Jimeno, Rita João, Vincent Johnson, Miriam Jordan, Reema Kanwar, Eva Kato, Sam Kebbell, Andrea Kennedy, Hubert Klum, Allan Kosmajac, Irene Kouroukis, Danica Kovacevic, Suzan Krepostman, Jory Krupse, Laurent La Gamba, Mary Lamorte, Joanna Lee, Ken Lee, Amandine Leriche, Dan Leung, Sarah Lewison, Avril Loreti, Brian Loube, Susana Lourenço Marques, Ana Luisa, Rodrigo Machado, Jaime Maddalena, Berry Mak, Nathaniel Mandigo, Helen Matloob, Matthew McClennon, Johnny McCormack, Jennifer McKnight, Roland McMahon, Darren Melrose, Bret Menezes, Neil Meredith, Kenneth Montague, Rodrigo Moreno, Sayil Moreno, Talya Moshinsky, Sara Moss, Kathleen Mullen, Cindy Munro, Michelle Murray, Meena Nanji, Laura Nanni, Maurice Narcis, Sarah Nasby, Harold Nelson, Fiona Ng, Niall McLaughlin Architects, Deborah Nolan, Angela Noussis, Neil O'Keeffe, Jacqueline Olivetti, Jamie Osborne, Roshanak Ostad, Peter Pallotta, Andres Pang, Joseph Paget, Annette Paiement, Andrea Parker, Stefano Pasquini, Mark Paterson, Jean-Michel Peers, Goran Petrovic, Terry Pidsadny, John Pinter, Krishna Persaud, Eva Prinz, John Quarterman, Ron Rada, Ramona Ramlochand, M P Ranjan, Christina Ray, Maxim Reider, Daniel Reiser, Marcia Ritz, Christiane Robbins, Nia Robyn, Frank Rodick, Ken Rose, Tom Rune Ostby, Umar Saeed, Eric Safyan, Liisa Salonen, Leah Sandals, Rita Saomarcos, Elizabeth Sarney, Galerie Saw, Tanya Scherbey, Diana Shearwood, David and Marianne Sidwell, Samantha Slicer, Marco Sousa Santos, Laura Splan, Vahagn Stepanian, Jessica Stuart-Crump, Surveillance Camera Players, Véronique Synnott, Dana Takeda, Danny Tan, Sebastien Terrean, Valerie Thai, Dan Thorpe, Din Tiranatvitayakul, Monica Tiulescu, Nancy Tong, Jacqueline Treolar, Gennadiy Tsygan, Florin Tudor, Alex Turner, Thomas Tsang, Claudia X. Valdes, Mona Vatamanu, Katie Varney, Tai Vo, Casper Voogt, Caleb Waldorf, Robert Walker, Ben Weeks, Karen White, Jenny Wilson, Andrew Yeung, June Yun, Nikolay Zanev, and Greg Zawidzki.

The support and, dare I repeat this, superhuman patience of our colleagues at Phaidon cannot be understated. Richard Schlagman, Karen Stein, and Megan McFarland have been through this odyssey with me twice now. Their encouragement has meant a lot to the development of this project, the Institute, and our studio. Caroline Green, our publicist at Phaidon, also deserves special thanks.

As I said at the beginning, Massive Change only exists because of the once-in-a-lifetime opportunities extended to us by the Vancouver Art Gallery and George Brown – Toronto City College. For the Art Gallery, aside from Bruce Grenville, we are especially grateful to its visionary director, Kathleen Bartels, for changing the rules about what is possible at a public art museum. And we must also thank the gallery's extraordinary team, including: Daina Augaitis, Jacqueline Gijssen, Bruce Wiedrick, Chris Wootten, Diane Robinson, Rosemary Nault, Marie Lopes, Colette Warburton, Karen Henry, Liz Zloklikovits, Suzana Barton, Susan Lavitt, Julie-Ann Backhouse, and Jenny Wilson. The gallery's brilliant public relations partner in New York, Resnicow Schroeder, have been invaluable in generating international media interest in the project – special thanks to Deborah Kirschner, Nathalie Hoch, and Fred Schroder.

At George Brown – Toronto City College, we are grateful to Paul Carder, former Dean of Business and Creative Arts, who provoked us to invent the Institute without Boundaries; his successor Maureen Loweth; and the Director of the School of Design, Luigi Ferrara, who all have been incredible supporters, as have several staff members there including Alice Lee, Eugene Harrigan, Michael Maynard, Ian Gregory, and Owen Pearce. Two members of George Brown's School of Design advisory board, Marlene Hore and Paul Rowan, encouraged the college to approach us in the first place and have continued to be partners with us on the project in a number of valuable ways. We especially thank Luigi Ferrara and the organizers of The Humane Village Congress (ICSID '97) for developing the phrase "Massive Change is not about the world of design; it's about the design of the world," which succinctly encapsulates the stance of our project.

The following must also be thanked for their support: Sam Ainsley, Lesley Beever, Ron Beever, Adam Berkowitz, Victoria Brown, Dan Cameron, Jill Cuthbertson, Nathalie de Briey, Rob Dickson, Kathryn Elder, Julie Fiala, Hamilton Fish, Robert Fitzpatrick, Gerry Flahive, Rod Fraser, Alexia Holt, Michelle Jacques, Nina Kaden Wright, Reiner Klein, Craig Leonard, David Leonard (in memory), Janet Leonard, Kathryn MacKay, Eleanor Marchand Dean Roger Martin, Trevor Maunder, Francis McKee, Scott McKenzie, Les Mendelbaum, Maris Mezulis, Jelena Mihajlovic, Linda Milrod, Miles Nadal, Chris Nanos, Alberta Nokes, Lisa Philips, Dennis Reid, Seona Reid, Michael Richer, Thomas Clifford Richer, Charlie Rose, Wolfgang Roters, Howard Saginur, Nicole Schneider, The Scottish Arts Council, Elizabeth Smith, Trevor Smith, David Suzuki, Matthew Teitelbaum, and Mike Zryd.

Lastly, I want to thank Bisi Williams, my wife, who has accompanied me on my journey into the unpredictable and thrilling terrain of Massive Change. She and our three beautiful daughters – Osunkemi, Omalola, and Adeshola – are my strength and inspiration.

Bruce Mau
July 2004

Editor's Note

I would like to express personal thanks to Bruce Mau, Jim Shedden, Cathy Jonasson, and Greg Van Alstyne of the Bruce Mau Design Studio and Ken Stowar and Brian Birchell of the University of Toronto's CIUT 89.5 FM for granting me the liberty to independently pursue the research and production that went into the design and development of Massive Change Radio.

Over the past year, I have sought out and had the honor of speaking in-depth with provocative thinkers across disciplines – and for that I am truly grateful. Thank you to all who took the time to respond to my emails and phone calls, and for agreeing to participate in the live-to-air or live-to-tape interviews. In order of appearance (from the first broadcast, August 26, 2003 to June 15, 2004), you are: Sir Martin Rees, Steve Squyres, Hazel Henderson, Alex Galloway, Nancy Padian, Dean Kamen, Leonard Shlain, Janine Benyus, Hernando de Soto, Fritjof Capra, Freeman Dyson, Ashok Gadgil, Michael McDonough, Rick Smalley, Bruce Sterling, Ian Foster, Seymour Melman, Matt Ridley, Bob Langer, Lawrence Lessig, Wendy Brawer, Patrick Moore, Felice Frankel, Andrew Zolli, Carol Burns, Stewart Brand, Philip Ball, Jeff Sachs, Bob Freling, Bill McDonough, Catherine Gray, Jaime Lerner, David Malin, Gwynne Dyer, James Der Derian, Arthur Kroker, John Broughton, Bill Drayton, Eugene Thacker, and Stephen Browne.

Sadly, due to space limitations, not all of you made it into the book – nor was I able, of course, to speak to all of the countless thought leaders out there that I wished to – but your brave new ideas have breathed life into the project nonetheless.

And, to my mind, this is just the tip of the iceberg in an effort to bring thoughtful reporting to the world and refocus the lens of journalism so that innovations and imaginings for the good of all make front-page news. Be not afraid of well-intentioned ambition and unstoppable conviction. These insights will carry us forward.

Jennifer Leonard
July 2004

Experts' Profiles

Carol Burns is a partner of Boston-based Taylor & Burns Architects. She has directed the Harvard Institute of Affordable Housing for twelve years, and has been a housing fellow at the Harvard Joint Center for Housing Studies. Carol received the AIA Education Honors Award for curriculum innovation in 1996 and in 1999 was awarded with the teaching prize by the Harvard Student Forum.

Michael McDonough is an architect and designer whose work has long explored the relationship between ecology and technology. He has designed commercial, residential, and urban planning projects, as well as objects and furniture. He is the author of *Malaparte: A House Like Me*, co-author of *The Smart House*, with James Grayson Trulove, and is currently working on a revisionist history of the Bauhaus.

Hernando de Soto is the President of the Institute for Liberty and Democracy in Peru, regarded by *The Economist* as the second most important think tank in the world. His mission is to work with heads of state and at the street level to redesign property law. He is the author of *The Mystery of Capital: Why Capitalism Triumphs in the West and Fails Everywhere Else* and *The Other Path: The Economic Answer to Terrorism*.

Dean Kamen is the President of DEKA Research and Development Corporation in Manchester, New Hampshire. He is an inventor devoted to human welfare and holds more than 150 patents for medical devices, such as the first drug infusion pump, and mobility devices, such as the Independence iBOT and the Segway Human Transporter. In 2002, he was awarded the Lemelson–MIT Prize for innovation.

Jaime Lerner is an architect, urban planner, and United Nations consultant for urban issues. He is the president of the Union of International Architects and the former (three-time) mayor of Curitiba, Brazil, and (two-time) governor of the state of Paraná, Brazil. Lerner led the urban revolution that put Curitiba – and, most notably, its public transit system – on the world map.

Robert Freling is the Executive Director of the Solar Electric Light Fund (SELF), a nonprofit charitable organization that helps rural villagers in developing countries improve their lives through clean, renewable energy and modern communications. He has been responsible for developing and coordinating solar rural electrification programs in Brazil, China, Indonesia, South Africa, Bhutan, Nigeria, Tanzania, and in the Solomon Islands.

Richard E. Smalley is the Gene and Norman Hackerman Professor of Chemistry and Professor of Physics & Astronomy at Rice University in Houston, Texas. His research in chemical physics has led to the discovery of a third elemental form of carbon ("fullerenes"), which is capable of being formed into a fiber 100 times stronger than steel at one-sixth the weight. He was awarded the 1996 Nobel Prize in Chemistry.

Lawrence Lessig is a professor of law and John A. Wilson Distinguished Faculty Scholar at Stanford Law School in Palo Alto, California. He teaches and writes in the areas of constitutional law, law and high technology, Internet regulation, comparative constitutional law, and the law of cyberspace. Lessig is the author of *The Future of Ideas, Code and Other Laws of Cyberspace*, and *Free Culture*.

Ian Foster is Associate Director in the Mathematics and Computer Science Division at Argonne National Laboratory and the Arthur Holly Compton Professor of Computer Science at the University of Chicago. He is the author of *The Sourcebook of Parallel Computing* and *The Grid 2: Blueprint for a New Computing Infrastructure*.

 Catherine Gray is the president of The Natural Step, an international advisory and research organization that works primarily with executives at Fortune 500 companies to help them integrate sustainability principles into their core strategy and operations. Gray has led watershed engagements with The Home Depot, McDonald's, and Bank of America, all of which have ongoing relationships with The Natural Step.

 Stewart Brand is the president of The Long Now Foundation and co-founder of the Long Bets Foundation. He founded The WELL (Whole Earth 'Lectronic Link) and founded, edited, and published the original *Whole Earth Catalog*. Brand works quarter-time as a consultant with Global Business Network and is the author of *The Clock of the Long Now, How Buildings Learn: What Happens After They're Built*, and *The Media Lab: Inventing the Future at MIT*.

 Hazel Henderson is a futurist, independent economist, and international consultant on sustainable development. She is the author of *Beyond Globalization* and seven other books. Her editorials appear in 27 languages in more than 400 newspapers, and her articles have appeared in over 250 journals, including *Harvard Business Review, New York Times*, and *Christian Science Monitor*.

 Felice Frankel is a research scientist in the School of Science at MIT and Director of the Envisioning Science Project. She collaborates with scientists to create compelling research images to better communicate ideas in science to the general public. She is the author of *Envisioning Science, the Design and Craft of the Science Image* and co-author of *On the Surface of Things*, with George Whitesides of Harvard.

 Philip Ball is a science writer and a consultant editor for physical sciences at *Nature*. His writings on science for the popular press have covered topical issues ranging from cosmology to the future of molecular biology. He is the author of eight books, including *The Ingredients, Stories of the Invisible, Made to Measure*, and *H_2O: A Biography of Water*.

 David Malin is an Adjunct Professor of Scientific Photography in the Faculty of Applied Science at Royal Melbourne Institute of Technology. Previously he was a photographic scientific-astronomer and a chemist, specializing in optical and electron microscopy, X-ray diffraction, and other techniques for exploring the very small. He is the scientific consultant to *Heaven & Earth* (Phaidon Press) and the author of numerous books and articles.

 Janine M. Benyus writes and educates in the natural sciences, teaches interpretive writing, lectures at the University of Montana, and works as a "biologist at the design table," helping designers, engineers, and business leaders consult life's genius in the creation of well-adapted designs. She is the author of six books including *Biomimicry: Innovation Inspired by Nature*.

Gwynne Dyer received his doctorate from the University of London in military and Middle Eastern history and for over twenty years has worked as a freelance journalist, columnist, broadcaster, and lecturer. His syndicated columns on international affairs appear in a dozen languages in nearly 200 newspapers published in more than 40 countries around the world.

 Seymour Melman is professor emeritus of industrial engineering at Columbia University in New York. He has long advocated the idea of an ordered transition from military to civilian production by military industries and facilities. He is the author of *After Capitalism*, *From Managerialism to Workplace Democracy*, *The Peace Race*, *Pentagon Capitalism*, *The Permanent War Economy*, *Profit Without Production*, and *The Demilitarized Society*.

 James H. Korris is the Creative Director of the Institute for Creative Technologies at the University of Southern California, which has undertaken numerous assignments relating to U.S. military transformation and the development of multiple software game projects for cognitive leadership training. The first of these, the X-Box squad-leader trainer Full Spectrum Warrior, won two E3 Critics Awards in June 2003.

 James Der Derian is Professor of Political Science at the University of Massachusetts, Amherst, and Professor of International Relations (Research) at Brown University, where he directs the Information Technology, War and Peace Project. His articles on war, technology, and the media have appeared in *The New York Times*, *The Nation*, *Washington Quarterly*, and *Wired*. He is the author of *On Diplomacy*, *Antidiplomacy*, and *Virtuous War*.

 Arthur Kroker is Canada Research Chair in Technology, Culture and Theory at the University of Victoria, Canada. A cultural theorist as well as co-editor of the prestigious electronic journal, *CTheory* (www.ctheory.net), he is author of numerous books on the future of technology and culture including, most recently, *The Will to Technology and the Culture of Nihilism: Heidegger, Nietzsche and Marx*.

 William McDonough is an architect and proponent of "The Next Industrial Revolution." He is the founding principal of William McDonough + Partners, Architecture and Community Design in Charlottesville, Virginia, and cofounder and principal, with Michael Braungart, of McDonough Braungart Design Chemistry, LLC, a product and systems development firm assisting client companies in implementing their unique sustaining design protocol.

 Bruce Sterling is a science fiction writer and a contributing editor to *Wired* magazine. He edited *Mirrorshades*, the definitive document of the cyberpunk movement, and co-authored the novel *The Difference Engine* with William Gibson. His most recent novel is *The Zenith Angle*. He lives in Austin, Texas, where he runs a weblog and a mailing list on environmental and design issues.

 Matt Ridley is the author of the international bestseller *Genome*, as well as *The Origins of Virtue*, *The Red Queen*, and *Nature via Nurture*. He has BA and DPhil degrees in zoology from Oxford University and was science editor and American editor of *The Economist* between 1983 and 1992. From 1996 to 2003, he was the founding chairman of International Centre for Life, the UK's first biotechnology village.

Freeman Dyson is a renowned mathematical physicist and professor emeritus at the Institute for Advanced Study in Princeton, New Jersey. He has worked in particle physics, condensed matter physics, astrophysics, nuclear engineering, pure mathematics, and biology. His celebrated books include *Disturbing the Universe, Infinite in All Directions, Origins of Life, From Eros to Gaia, Imagined Worlds*, and *The Sun, the Genome, and the Internet*.

Robert S. Langer is the Kenneth J. Germeshausen Professor of Chemical and Biomedical Engineering at MIT. He has over 500 issued or pending patents worldwide and has received over 120 major awards, including the Lemelson–MIT prize for medical innovation. In 2002, he received the Charles Stark Draper Prize, considered the equivalent of the Nobel Prize for engineers, from the National Academy of Engineering.

Eugene Thacker is Assistant Professor in the School of Literature, Communication, and Culture at the Georgia Institute of Technology. He has written extensively on the cultural, social, and political aspects of biotechnologies, and is the author of *Biomedia* and *The Global Genome: Biotechnology, Politics, and Culture*. He also works with Biotech Hobbyist.

Ashok Gadgil is a senior staff scientist and group leader in the Environmental Energy Technologies Division, in the Indoor Environment Program, of the U.S. Department of Energy's Lawrence Berkeley National Laboratory. Gadgil's invention, the UV Waterworks, received *Discover* magazine's 1996 Award for Technological Innovation (Environment Category) and *Popular Science* magazine's 1996 Best of What's New Award.

Bill Drayton, a former McKinsey & Co. consultant, founded Ashoka: Innovators for the Public in 1980. He briefly served in the White House and has taught both law and management at Stanford Law School and Harvard's Kennedy School of Government. Drayton was elected a MacArthur Fellow and was awarded the Yale School of Management's Award for Entrepreneurial Excellence.

Stephen Browne is Director of the Information and Communications Technology for Development Group of the United Nations Development Program. He has spent some 16 years in fieldwork for the UN, including assignments in Thailand and Somalia, and – as UN Resident Coordinator – in Ukraine and Rwanda. In 1999, he became Director of the Poverty Reduction Programme for UNDP and moved to his present position at the end of 2002.

Nancy Padian is an epidemiologist, internationally recognized expert in the heterosexual transmission of HIV, and a professor in the Department of Obstetrics, Gynecology, and Reproductive Sciences at the University of California San Francisco. In 1994, she co-founded the University of Zimbabwe–UCSF Collaborative Research Program in Women's Health in Zimbabwe, and in 2001 she founded the UCSF Women's Global Health Imperative.

Jeffrey D. Sachs is the Director of The Earth Institute, Quetelet Professor of Sustainable Development, and Professor of Health Policy and Management at Columbia University. He is Special Advisor to UN Secretary General Kofi Annan on poverty reduction initiatives called the Millennium Development Goals and is renowned for his work with international agencies to promote poverty reduction, disease control, and debt reduction of poor countries.

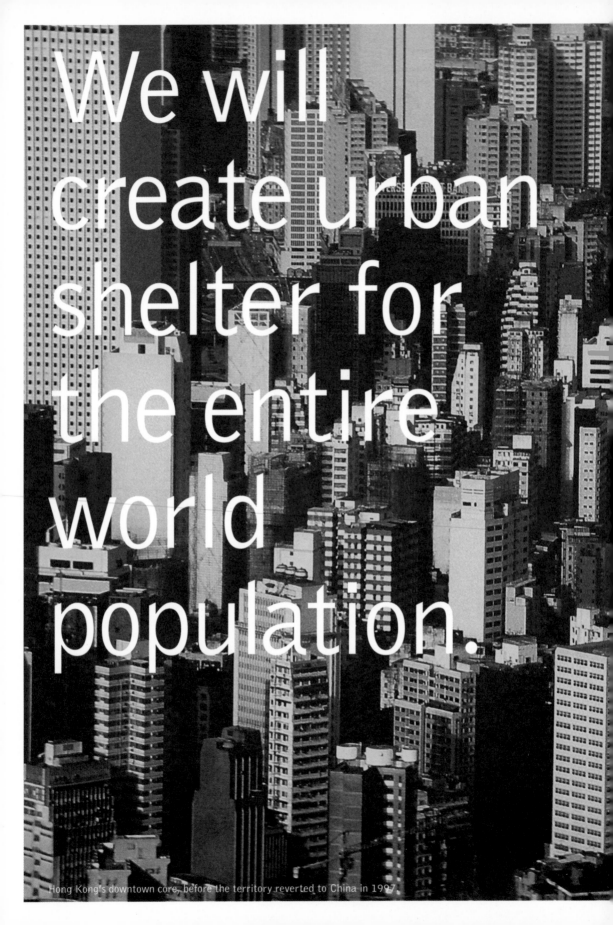

We will create urban shelter for the entire world population.

Hong Kong's downtown core, before the territory reverted to China in 1997.

Urbanization, one of humankind's most successful and ambitious programs, is the triumph of the unnatural over the natural, the grid over the organic. We remain committed to a global program of extrusion upward and repetition outward in an effort to provide shelter that is safe, healthy, and uplifting. Underway on a scale never before witnessed, one side effect of urbanization is the liberation of vast depopulated territories for the efficient production of "nature."

URBAN ECONOMIES

1.01

1.02

1.01 DYnamic-MAXimum-TensION. Buckminster Fuller's ingenious Dymaxion Dwelling Machine enclosed a maximum amount of space with a minimum amount of material and expense. Conceived and designed in the 1920s, it was Fuller's solution to the need for a mass-produced, affordable, easily transportable, and environmentally efficient house.

1.02 BABY, IT'S WARM INSIDE. Manufactured and cut to order, SIPs (Structural Insulated Panels) are high-performance building materials that reduce site labor and provide insulative value to homes in harsh climates, like Antarctica. The tight building envelope these sandwich panels create allows for more than 50% energy efficiency.

1.03 PRE-FABULOUS. With its roots in the pre-WWII Sears & Roebuck mail-order home strategy, IKEA and Skanska partnered in Sweden in 1996 to build BoKlok ("live sensibly") homes for small households with limited financial resources. There are now more than forty-five constructions in Sweden and the concept has been launched and built in Finland, Norway, and Denmark.

1.03

Manufactured housing: In one century the world population has gone from one to six billion, while life expectancy has doubled. The problems we share are plural. Architectural practice and education, however, are still locked to the idea of the singular.

The manufactured house is the brick of the twenty-first century.
– Paul Rudolph (1918–1997), American architect

There are extraordinary examples of synthesis between manufacturing and architecture – historically, and in place today – where urban development has taken advantage of industrial technologies and efficiencies, with a view to make affordable and easy-to-assemble domestic spaces. As a result, we've witnessed over time that the potential exists to meet housing needs around the world.

There has been a sort of global good citizenship at play since the very early days of manufactured housing. As far back as the 1830s in London, efforts were made to house homeless immigrants arriving in England; then came Buckminster Fuller and a long line of architects curious about industrial processes, including, and certainly not limited to, Jean Prouvé, Archigram, Moshe Safdie, Paul Rudolph, Deborah Berke, and Future Systems.

It's not surprising that the practice of architecture as a whole has not been involved with the design of manufactured housing – blame it on the cheap appearance or tendency to cluster them in trailer park ghettos because of zoning regulations. Nevertheless, the American public has embraced it wholeheartedly: mobile homes already shelter more than 12.5 million Americans.

The manufactured house also has myriad potentials in terms of waste reduction (30% of the drywall that arrives at a construction site leaves in the dumpster) and the challenges that come with dense urban areas. These potentials are only now being fully explored.

Carol Burns
on manufactured
housing

What is the true definition of a manufactured house?
At the bottom line, what defines a manufactured house is a permanent chassis. This is a fundamental physical feature by which we know this house type that has gone under various names at other times – the travel trailer, the house trailer, and, for a long period of time, the mobile home. Ever since the HUD [U.S. Department of Housing and Urban Development] Code was put into effect in 1974, it has been called a manufactured house. The funny thing is, it doesn't mean anything out of a factory; it means precisely a house that moves down the highway on its own permanent chassis. It's a hybrid entity, having as much to do with travel trailers and transportation vehicles as it does with housing.

When did the notion of mobility enter the realm of housing?
Issues of mobility have influenced dwellings even perhaps before settlement – the nomads were mobile dwellers, of course. But in terms of indus-trialization, which is the more relevant epistemo-logical era of considering this, the U.S. railroad system allowed prefabricated buildings to be shipped around the country. In the popular summer destination of Martha's Vineyard, for example, there are Queen Anne gingerbread-style houses that were built just after the Civil War, originally ordered from the Sears catalogue. They weren't aimed at the mobility of the people, but they

were made available to this island by way of a transportation infrastructure.

When did the design of trailers change to reflect their new role as a dwelling space?
When the federal government had to move soldiers and their families around the country during WWII, the house trailer fit the bill. They required fast construction of large-scale settlements, and although the house trailer could be moved, it had more interest for its occupants in a set location. In the subsequent peacetime economy, there was such a pent-up demand for housing that trailers as permanent homes became more widely avail-able and more acceptable.

Were there any significant housing initiatives that took advantage of the manufacturing technology immediately after WWII?
The story of Levittown is the best example of retooling factories for civilian purposes in post-war America – the "swords into plowshares" idea. [William] Levitt began to develop housing at a scale and in a way that had not been seen before. It's the way that we understand the development of suburbs now, but at the time it was a radical idea. Levitt referred to it as a "factory under the sky."

What sort of aesthetic patterns have you come to observe in your research?
Over 95% of American housing construction is in the detached home – it seems to be a cultural preference that's not easily denied. Its alluring features are private outdoor space and a place to park the car nearby, if not directly in the house. Also, in these changing times, we as a society have overwhelming preferences toward aesthetic expressions of the past. People depend on the physical environment to provide them with a

1.04 Aerial view of Levittown, Pennsylvania, 1959.

certain kind of continuity or "lasting," as [Spanish architect] Rafael Moneo would call it. With that comes an over-sentimentalized desire to have even new things look old.

Are there manufactured solutions for both the suburban and increasingly dense urban environments?
Local codes have specified in the past that manufactured housing be located in less desirable areas – the other side of the tracks or low-lying land, down by the river. With the rise of zoning as a primary tool of planning, it's often the case that manufactured homes exist as a separate homogenous entity. But it need not be the case. In the urban context, the manufactured house is morphologically a great candidate for infilling empty lots that were developed before WWII, whose configuration is deep, with a short side facing the street.

Why is manufactured housing an attractive area of investigation for you?
The fact that this house type comprises the lowest rung on the stepladder of home ownership makes it so inherently interesting to me. Its low price is due to the efficiency of factory production, the closed conditions, the twelve-month work cycle, and the fact that it greatly reduces construction waste. It's true that manufactured housing too often has looked cheap and has been stigmatized in various locations because of it. But I believe good design need not cost more than poor or inattentive design. I'm interested in exploring this further.

You refer to Karl Popper's notions of clocklike and cloudlike systems in relation to industrialized housing. What's the connection?
Karl Popper was a theoretician who was interested in the processes of early industrialization and anticipated some of what has been characterized much more recently as chaos theory. He recognized that some processes are like clocks – engineered, linear – and some processes are like clouds (or gnats or birds) – less predictable, more open-ended. With respect to manufactured housing, many of the early efforts by [Buckminster] Fuller, [Jean] Prouvé, [Konrad] Wachsmann, etc. wanted to work in line with the industrialized process, and so they tried to make "clocks." When the systems at play in the field had nothing to do with "the

clock" that they were designing, their projects ground to a halt. Popper suggested an alternative way of thinking, something more amorphous that could be harnessed to understand the processes at play, which many industries are getting at today by way of mass customization. Mass customization allows for a high degree of individuation and a certain kind of unpredictability in the line of production that takes advantage of efficiencies without necessarily producing the same thing everywhere.

Do you think Buckminster Fuller's prediction of housing becoming a service rather than a commodity will ever come true?
One way that I've understood this idea of his has to do with an assessment of what should be provided to everyone. For instance, we look at

> It's true that manufactured housing too often has looked cheap and has been stigmatized in various locations because of it. But I believe good design need not cost more than poor or inattentive design.

utilities like water as a service – and there's a regulated market that's constructed to provide equally to all consumers within the market. Housing has never been in such a category in America. The American real estate system looks at houses as investments. The possibility of saving and gaining based on home ownership is one of the most important financial opportunities that most families in this country face. So I can understand the benefit of thinking about housing as if it were conceived of as a service, but it's hard to imagine the system that we live in undergoing such a transformation. I think the Buckminster Fuller idea was a great polemical challenge and, like other early visionaries of industrial processes, he has inspired many of us to search for possible futures.

Carol Burns is a partner of Taylor & Burns Architects in Boston, Massachusetts.

Density offers hope: With nearly half of the world's population living in cities, density is increasingly becoming the global condition. The denser we make our cities, the more we can sustain ecosystems.

We need to draw lines in the ground and say, "The concrete stops here." That forces people to build in and up, rather than out – and there's nothing wrong with high, dense urban environments as long as they're planned correctly. They can be extremely livable. They tend to require less transportation, fewer sewer lines, fewer power lines, fewer roads, and more tightly packed structures, which in and of themselves are more energy efficient.

– Patrick Moore, Greenpeace cofounder and environmental consultant

The more reasonably we can fit together, the more we can support what surrounds us. If each member of our current global population (6,376,394,000 and growing) lived in his or her own Levittown house (750 sq. ft.), then we would use approximately 440,085.8 km² of land for housing. Surprisingly, this is only 0.35% of all the Earth's 126,909,000 square kilometers of habitable land. Factoring in the average density of urban "sprawl" – three homes per residential acre – we would still inhabit only 6.8% of all the Earth's habitable land. (Add to this the massive space needed to accommodate highways, farmland, city centers, and all other infrastructure required to support our global population.)

Now imagine if we stacked the world's population vertically. We'd take up dramatically less space. As the world urbanizes, we need to acknowledge the ever-growing condition of urban sprawl, design it better, and simultaneously look to density as a solution, which will liberate territory for the production of nature. If people live in cities, they don't destroy the country. As we create density, we simultaneously open up the surrounding rural space.

DENSE CITIES. Population per square kilometer: Chicago (1.05): 1,511; New York (1.06): 1,728; Tokyo (1.07): 5,934; São Paulo (1.08): 8,378; Cairo (1.09): 25,325; Hong Kong (1.10): 49,581.

Michael McDonough on sustainable architecture

Your e-House is a unique mix of high performance and alternative technologies. What was your inspiration?
I was prompted by an article I did with Bruce Sterling for *Wired* called "Newer New York," which became a focal point for learning about and consolidating information on all the building products in existence. Much to my surprise, every single thing I could think of – if I Googled around enough – I found. And more often than not, I discovered that I could buy it with a credit card and have it shipped overnight to a building site. I quite literally found the future of building on the Internet, waiting to be purchased and implemented. This notion became the basis for a science fiction story, but then I started thinking about building it for real. My wife and I were looking at property in upstate New York at the time, so we decided to go for it.

How is e-House a metaphor for the community?
Buildings should not be considered as isolated objects. It's profoundly important to understand how they're connected to the ground and the sky, and how they're connected to the culture of an area. In terms of technological connections, or connections to the culture of technology, this means making a building that thinks for itself, analogous to the way a human body functions. I'd like the building to adjust itself according to temperature and send email alerts when it needs attention.

How is holistic thinking important to architecture?
Every building has connections to the sky, ground, and community, but these could be appreciated and utilized much better. In e-House, we collect rainwater to irrigate our garden. We also use it to store energy from sunlight and earth, and that energy is used to heat or cool a hyper-energy-efficient house. If you extend this thinking to other building systems, you can engineer a geothermal field for maximum efficiency by backfilling it with clean, well-drained, fertile soil, and get both a heating and cooling source for your home and a productive organic garden. The more people start doing this community-wide, the more open space and forest can be conserved. This, of course, is an alternative to suburban sprawl. If government encourages this tendency through tax policy, you get large organic districts with hyper-energy-efficient homes.

Such districts can have economic and social value. We planned e-House to have an organic microfarm, greenhouse, and agro-forestry (this is located in New York City's watershed – a 1900-square-mile district that feeds the city's reservoirs, delivering a billion gallons of potable water daily). Imagine that new home-building in this vast area was encouraged to have organic microagricultural uses. New York City and its surrounding areas would be tethered to each other – clean, pure water from organic watersheds and urban markets for local organic produce. So the land and buildings can multitask and form mutually beneficial relationships at any scale. This is the sort of productive, holistic thinking I want to encourage in architecture and, in turn, in public policy and regional planning.

How are megacities impacting the environment?
There's nothing to say that cities are in and of themselves bad. You can do green buildings, you can do green infrastructure systems, and you can reduce vehicular transport by using mass transportation. You can consolidate building construction and heating and cooling systems in larger structures – not that you should do megastructures, necessarily, but there are efficiencies of scale. The surprise is that New York City is one of the most environmentally efficient cities on the surface of the planet because of its density. So density is not necessarily the enemy. It seems that suburban sprawl is the enemy; this is where you need to have more profound innovation. My suggestion is that you stop thinking about the city and the suburbs and the exurbs and the rural areas as separate entities and you really consider them as united ecosystems.

From a design perspective, are there exemplary building materials for dense environments?
I like bamboo a lot. The more you use it, the better things get. It's deeply versed in cultures all over the world, it's stronger than steel in tension, it's stronger than concrete in compression, and it's more stable than red oak, which is a very stable flooring. When you plant it, it acts as a

adjusted regionally. At the same time, you can have almost infinite design flexibility, which is a problem, for example, with modular houses.

What does it mean today for designers to think globally?
Well, I think whenever you sit down at the drafting board or the computer, whatever your media might be, it's important to ask yourself what the implications of a project are, beyond the object itself – whether the object is a building or a chair or a table. To constantly ask, "What are the processes by which this thing is produced?" and "What are the processes by which it will end its useful life?" Beyond that, I like the idea of multitier thinking. It's always good to investigate the possibility of making any one system do multiple things. As I mentioned with the e-House, if you're going to spend a lot

> The surprise is that New York City is one of the most environmentally efficient cities on the surface of the planet because of its density. So density is not necessarily the enemy. It seems that suburban sprawl is the enemy; this is where you need to have more profound innovation.

bioabsorber, cleaning pollutants out of the soil; it simultaneously stabilizes the soil and prevents erosion. While it's doing all of these good things, it returns more oxygen to the air through photosynthesis than almost any other deciduous plant.

of time and energy engineering one of its systems, shouldn't you also ask yourself what else that one system could do?

Michael McDonough is an architect in New York.

It's clear, then, we need to plant more bamboo around the world!
If we did that, it could have a profound effect on carbon sequestration, which means it could sequester carbon out of the atmosphere and deal with global warming related to the problem of greenhouse gases.

Do you think architecture can link up with other disciplines, like manufacturing, to start addressing the global housing crisis?
Architects continue to attempt universal solutions. The problem is, we don't have a universal climate. And you really need to adjust your building types to climate. What makes sense in southern California may not make sense in northern Canada, for example. But there are useful manufactured components of buildings, such as structural insulated panels that have high insulative value. You can have a design produced on a computer and email it to one of the manufacturers, and they will cut out the doors and windows and number the panels and put it together like some sort of fanciful Lego system. This allows for a distributed manufacturing system that can be

1.11 e-House, 2000.

1.12

1.13

Expensive real estate in Tokyo (1.12) versus homes
made by hand in Ankara, Turkey (1.13). In organic
urban development, there is no grid; there is only
spontaneous juxtaposition and the constant negotiation of boundaries.

The entrepreneurial Third World: By necessity, the Third World consists of entrepreneurs who struggle to build informal urban environments by hand. What if we were to design a property system that supports this?

The architect is sort of this almighty figure who dictates, subjugates reality to number and calculation and construction and result. But for some reason, every ten to thirty years this issue of informality comes back. The struggle of people building their own environments is the ultimate Utopian idea.

– Teddy Cruz, San Diego–based architect

When we look to law and how it's being changed in response to the massive changes on a global scale, it's useful to consider the work of Peruvian economist Hernando de Soto. De Soto works toward redesigning property law around the world with the aim of granting property rights to the poor of developing and former communist nations.

We tend to think that if people have a roof, they have shelter. De Soto says you need more than a roof. You need that roof in the system. And it's the system that provides the potential for the poor to increase wealth.

He and his team at the Institute for Liberty and Democracy in Lima gathered "dead capital" data in Egypt, Haiti, Honduras, and Mexico – that is, they calculated "the percentage of assets that can only be used as shelter or business tools, but not as a means to obtain collateral for a loan, to generate investment, or to create additional functions to obtain surplus value." The numbers are astounding: The portion of the total population holding its real estate assets outside the legal system in Egypt is 92%, in Haiti 82%, in Honduras 86%, and in Mexico 80%.

There's an incredible untapped entrepreneurial force that is hard at work outside the system. Imagine the effect if these homes and businesses were accounted for!

Hernando de Soto on redesigning property law

How do your ideas differ from conventional ideas on the source of capital?
I would say that they add to conventional ideas rather than differ from them. I believe that the only way we're able to capture capital is through property. Property is the system that allows us to capture the value we have in things in a manner that's tangible and that allows us to transfer value so as to initiate new causes and sources of growth.

T. S. Eliot asks, "Where is the wisdom we have lost in knowledge and where is the knowledge we have lost in information?" In terms of capital, this is an interesting metaphor.
Right. The information contained in the property system – or concretely in the recording system, a deed, a share, or something that represents equity – represents its value. The value is captured in the property document. In this sense, property is a system of representations of value. If you don't have a property system, you don't have the representational devices with which you can then capture value, store it, make it liquid, and invest it.

The majority of developing and former communist countries do not have property systems that allow them to concretize the value of the many things that they produce and own. As a result, they are unable to grow. So they end up rejecting the system. Interestingly, even developed countries are not conscious of what makes the accumulation of capital possible, despite the advanced stages of their property systems.

How does your Latin American perspective shape your understanding of global economics?
It allows me to understand that there are things more concrete than cultural tracts that keep people from advancing. There are major legal obstacles to bringing people together to produce efficiently and to accumulating and using capital. In North America, you see people from different walks of life with no prior relationship to each other getting together because it's the right combination of people for producing something in particular. They need not know each other to fit together well, at a technical and economic level, because all the legal means to cooperate and accumulate capital are in place. In countries like Peru, they're not.

One day, you and your research team at the Institute for Liberty and Democracy closed your books and opened your eyes. Tell me about those early days of going deep into the world's poorest communities.
When the books weren't working for me anymore, I decided in the early 1980s to visit the shantytowns of Lima to see what was really going on. I came across an area with paved roads and buildings made of solid brick and iron on one side and a mess of cardboard and old tires on the other side. What struck me was the fact that both sides were actually the same shantytown; they had been created at the same time when people from the same Andean village migrated there. But when the road was paved, it split them in two, and then they each elected a different leader and took a different course. But they started on an even basis.

The difference was that those in the more developed section had obtained titles to their buildings and those that had small enterprises had secured licenses to function. On the less developed side, they had used their time to obtain material things, but nothing was legalized. That's when I started saying, "Oh my God, the law does seem to make an enormous difference." I understood then that those who had titles and a license to operate could, all of a sudden, count on different devices that allowed them to use their homes and businesses as a guarantee against loans so they could start getting credit.

As soon as there's legally accredited property in place, property documents can be used to create a variety of functions for their owners that the assets themselves cannot. People are able to meet in the marketplace of representations and relate to each other productively.

Does capital begin to work only when one has a personal investment in something?
I think it was President Clinton's Treasury Secretary Larry Summers, who's now president of Harvard, who said, "Nobody's ever washed a rented car." So, yes, the moment you own something you care more for it. But I'm also saying that, beyond ownership, being within a comprehensive property system makes possible a series

Monetary Fund] would come around and pin a medal on our chest. Fujimori gave us less leeway to do what we thought was really important – this was one of the reasons that we split with Fujimori. Macroeconomics is simply unsustainable over time unless you also have the micro: property networks and capital-creating systems that underpin it and make even the poorest participate in the social contract that it rests on.

You've been an adviser to Egypt, Mexico, and, of course, Peru. Where else are you implementing change in the world?
We started programs in the Philippines and Honduras and we're about to start 21 new ones – from Russia to Tanzania, Nigeria, Ghana, Kazakhstan, and Georgia. We should be able – if we can build up our capacity – to help all these heads of states and be in 20- to 30-odd countries by 2005.

The information contained in the property system represents its value. The value is captured in the property document. In this sense, property is a system of representations of value. If you don't have a property system, you don't have the representational devices with which you can then capture value, store it, make it liquid, and invest it.

When you were inducted into the International Democracy Hall of Fame, what did you have to say about economic development?
I said that the law and how you build it is crucial. Not how much you know about law, but the genesis of law, how it starts up. Westerners many times forget how law begins because the system is already in place. I'm not one of these guys who believes that the West is guilty for not teaching us. I think this is more about internal politics than about foreign aid. It's our responsibility first. But since the West does want to help, they should know that the law is much more important than many other items and that there is still no adequate support for it in international organizations.

of things that were not possible before. For example, investment. Nobody is ever going to invest without obtaining a share (a legal creation) that bears witness to his/her investment. Property does more than certify ownership; it's an incentive. An integrated property system is the mother of many other institutions – it makes credit systems possible; no, more, it allows police to "skip trace," locate criminals, and make people accountable for their actions.

When you worked with former president of Peru Alberto Fujimori, how did you discover that micro-, not macro-, economics is the way to bring about massive change to the world's poor?
Although we knew from the very beginning that we had to get into microeconomics – that is, empower people with property – we were more successful with macro. We didn't have many apostles throughout the world doing this kind of work at the time. When we did the macro, all heads of state would applaud, big businessmen would applaud, the IMF [International

Hernando de Soto is president of the Institute for Liberty and Democracy in Lima, Peru.

1.14 CLEAR FROM ABOVE. NASA's Terra satellite was able to capture this image of the U.S.–Mexico border in California. Each segment is its own micro-managed system, yet all are alike in that they are regulated and designed.

Everywhere is city: We still conceive of cities as discrete objects, separate from their surroundings. That's no longer true. There is no exterior to the global city that connects and sustains us all.

One thing is sure. The earth is now more cultivated and developed than ever before. There is more farming with pure force, swamps are drying up, and cities are springing up on unprecedented scale. We've become a burden to our planet. Resources are becoming scarce, and soon nature will no longer be able to satisfy our needs.

– Quintus Septimus Florens Tertullianus, Roman theologian, 200 B.C.

Since the dawn of agriculture 10,000 years ago, the human tendency has been to manage land. Cities evolved in a defensive posture, an inside protected against an outside. More and more, we're embracing the stewardship role and increasing and extending the level of management. We must extend design and stewardship to encompass all terrain. The new global city is now defined with zones of urban, suburban, rural, leisure, and even "natural" precincts – all managed, all part of a designed system.

Instead of isolated parcels of land or singular architectural projects, it's now a matter of considering an entire city infrastructure and its connected environs, whose reach is hundreds of miles beyond what has been conventionally considered urban domain. The city now represents all territory, and all territory needs to be regarded and managed as one urban system.

The contradiction embodied in the practice of architecture is that it has traditionally chosen to focus on big buildings rather than to see the big picture as the most compelling design project. Architects have tended to build pieces of city without regarding their relationship to the whole. But holistic thinking is exactly what we need here if we're ever to develop the capacity we need to provide shelter on a global scale.

It's clear that synthesis is not merely useful: it's critical.

We will enable sustainable mobility.

Maglev test facility in Emsland, Germany.

New design developments create a synthesis among energy, manufacturing, computing, and materials that promises to revolutionize movement. From short-distance personal travel to supersonic global tourism and the transport of massive payloads, our new economies of movement are reconfiguring the urban and colonizing what remains of the natural terrain.

MOVEMENT ECONOMIES

2.01 RELIEVE CONGESTION. With the electrically-powered Segway Human Transporter (HT), Dean Kamen's dream is to reconfigure dense urban environment and accommodate a cheaper, cleaner and more efficient alternative to the car. According to him, due to traffic congestion, 43% of our fuel is used while we're sitting still.

- Road traffic injuries are predicted to become the third-largest contributor to the global burden of disease by 2020

- Road traffic deaths are predicted to increase by 83% in low-income and middle-income countries, and to decrease by 27% in high-income countries; these figures amount to a predicted global increase of 67% by 2020

- It is estimated that every year road traffic crashes cost US$518 billion globally

- The average annual traffic delay per person in the United States has climbed from 11 hours in 1982 to 36 hours in 1999

Personal freedom: The world hasn't embraced secular democracy, but it has embraced traffic. The radical success of the car has brought about its failure. Personal mobility projects are under way worldwide to deliver maximum freedom with minimal impact.

The twentieth century is not just a century that has automobiles. It's also a century that doesn't have horses. This is an insight that we need to get our heads around in some basic way. We need to overcome this blindness that we've had about what the process of time does to us.

– Bruce Sterling, science-fiction writer

The car eliminated the problems associated with the horse and buggy and answered the need for personal liberty. But its success brought about a new set of problems. With millions of cars now clogging up the urban landscape in both the developed and developing worlds, the global design challenge is to dream up lighter, smarter, and less expensive options.

From the motorized scooter and Sinclair C5 to the Moller Skycar and multimodal, task-appropriate transportation options and systems developed by CarboyMobility (headed by Dan Sturges), personal freedom is not about a single technology or project. It's really about a vector in the scatterplot of ongoing projects whose goal is to produce liberty with greater efficiency and less environmental impact.

New developments in the realms of energy and manufacturing promise to revolutionize short-distance surface mobility. In particular, inventor Dean Kamen's Segway Human Transporter (HT) could very well "shrink" cities and make cars cumbersome. The challenge to overcome is our cultural attachment to a four-wheeled dream machine and the infrastructure that supports it.

2.02 FROM PROTOTYPE TO COMMERCIAL PRODUCT. The Segway HT is more than industrial design. It is a synthesis of innovation in computing, gyroscopic sensing, energy, materials, and manufacturing. The result is not only a new product, but also a new product category. Succeed or fail, the idea of the Segway HT is here to stay.

"Consider the following: a 125-pound woman drives a new Toyota Sequoia (SUV) down to the nearby commercial center to drop off the videos and pick up a coffee. The Toyota SUV weighs forty-four times her weight. So imagine the way royalty were carried on a throne in the ancient times by people. Imagine her being carried by forty-four of herself on a platform. The forty-four of her are carrying her to take the videos back and to pick up the coffee. A Segway is 60% of her weight. I think this helps to visualize the absurdity of it all."

– Dan Sturges, new mobility designer

2.03

2.04

2.03 "4-D TWIN, ANGULARLY-ORIENTABLE, INDIVIDUALLY-THROTTLEABLE, JET-STILT, CONTROLLED-PLUMMETING TRANSPORT." Buckminster Fuller wrote the highfalutin description above to describe his three-wheel Dymaxion Car invention in 1927. This model traveled 120 miles per hour, was 19 feet long, and could turn on a dime.

2.04 C5 GOING ON SIX. Originally launched in England by Sir Clive Sinclair in 1985, the C5 was a commercial flop, with only 12,000 units sold. But Sir Clive remains committed to the project and plans to soon release an upgraded model, the Sinclair C6.

2.05

2.05 LIGHTEN UP. Cars need to get lighter before they can be fuel efficient. As a result, Hypercar, Inc. (now called Fiberforge) directs its research efforts toward developing lightweight, safe composite auto bodies. The "Revolution" is based on the 2003 World Technology Award–winning Hypercar concept by Amory Lovins, co-founder of the Rocky Mountain Institute.

•

2.06 READY TO POP. Montreal-based Bombardier Recreational Products asked, "What will the future of recreational transportation be like?" Their answer was a hydrogen fuel cell-powered, gyroscopically balanced, one-wheeled recreational and commuting vehicle concept called the Embrio. No working model has hit the market yet.

2.06

2.07 SURF'S UP. Already in production and available now from Brazil is the JetWheel from Wheelsurf Sport Ltda., another unicycle experiment, but with this one its rider cruises around inside the wheel. The developers say it allows for the "inner circle experience."

2.07

2.08 JUST RIGHT. There's no such thing as one-size-fits-all when it comes to getting around. Once this sentiment is widely accepted, so too will be subcars, a category of vehicles between the scooter and the car. Nevco's Gizmo "fits into places other cars don't, like the ecosystem." A great student vehicle that does what scooters can't in unsavory weather, it comes with the following advice: "Instead of a bigger car, try a bigger idea."

2.08

2.09 CARS ARE GREAT, BUT... In an effort to address the gap in personal mobility options between the most inexpensive motor scooter and smallest of cars, Dan Sturges designed the first Neighborhood Electric Vehicle (NEV), now called the Global Electric Motorcar (GEM). It functions as the subcar component in a network of options, which includes public transit. Sturges admits, "Cars are great, but the automobile monoculture we've created is not. Every major urban area today is working to enhance options to car use."

2.09

Dean Kamen
on personal mobility

The Segway Human Transporter (HT) is efficient where urban driving currently is not. How so?
Cars are perfect machines for a discreet mission. Their performance is optimized at 50, 60, 70 miles an hour. They carry you and your whole family and keep you warm in the winter and cool in the summer and move you between cities. It's fun to do that, it's efficient to do that. Then you put a car in the middle of a city, with one person in it. That one person is trying to get one mile or two at a speed much less than the distance he or she wants to travel. It's absurd. Especially since half of all car trips in the U.S. are less than a few miles. With cars creeping along causing congestion and pollution, they are no longer efficient. And with half of the human population now living in cities, there has to be another way to travel short distances. What if you could give the pedestrian, with a Segway HT, the ability to glide along the sidewalk at 8, 9, or 10 miles an hour? You would have given them a safe option that is not only cleaner, but also more efficient. And, by the way, a lot of fun!

Dean, what is it about a Segway HT that makes people smile when they step on it?
When you first climb aboard a Segway HT, you feel like a kid when he stands up for the first time. You're bewildered. As you think forward, the machine starts to move forward. As you feel yourself wanting to step back, it intuits this and reverses. The experience of being on a Segway HT isn't really like anything else. It's like watching yourself learn to acquire the capability to balance.

Seeing that there are no brakes, engine, throttle, gearshift or steering wheel, I assume you would agree with Arthur C. Clarke's statement, "Any sufficiently advanced technology is indistinguishable from magic."
I not only agree with that statement, I would say that we work hard to live up to that standard with most of our projects. If a client doesn't say, "Wow, that's amazing!" I assume we haven't succeeded yet. Building the iBOT 3000 Mobility System, an enhanced advance on the wheelchair that uses cutting-edge robotics to allow the disabled to stand up, look their colleagues in the eye and walk up a flight of stairs is one very vivid example. Building medical equipment so that

people can dialyze themselves at home instead of living three nights of every week in an iron lung in a hospital room is another. We dedicate ourselves to projects that significantly change people's lives.

How do urban centers in both the developed and developing worlds need to be "rearchitected" to accommodate the Segway HT?
If you wanted to put a Segway HT in ancient Greece or modern-day megacities like Tokyo, New York, Paris, or Mexico City, the only thing you have to do is change people's attitudes. Transportation infrastructure is not required. If people can walk there, Segway HTs can roll there. To the extent that you allow this to happen, every time a Segway HT takes one of those trips, it eliminates the need for a car. It frees up the space. It cleans up the air. Everybody wins. The Segway HT is an option that doesn't require we fill up the spaces between buildings with 3,000-pound machines snorting at each other. Cities were meant to be pedestrian environments, with people moving

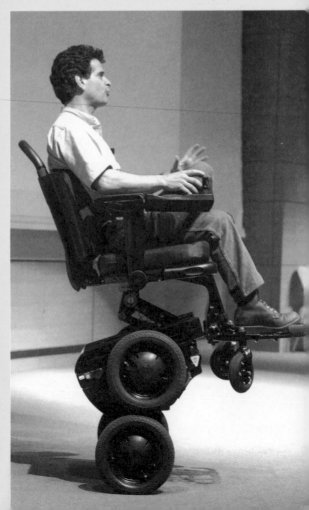

2.10 Dean Kamen on his iBOT 3000 Mobility System.

shoulder-to-shoulder through the space. Let's take them back and make them that way.

Has the automotive industry tried to hold you back at all?

When it comes right down to it, ours are noncompetitive technologies. Cars and planes aren't really competitive either. A plane is a great way to travel thousands of miles, the car is good for 50 or 60 miles, but when you get to that last mile or two the plane is totally absurd and the car is almost as absurd. So I think the automotive industry recognizes that the Segway HT is what it is – one of many mobility alternatives.

I understand that you have quite an extensive and impressive collection of vehicles.

Yes, it's true. I am not a naïve green person who insists on eliminating technology. I'd be the first to admit that I have cars and helicopters and airplanes. When the machine fits the mission, I believe that technologies improve the quality of our lives. The problem with this last mile – the niche distance between walking and driving – is that nobody until now had the right technology to apply to it.

Both the iBOT and the Segway HT employ complex software and microprocessor technology. But they also carry the wisdom of the gyroscope. Why is that?

The physics of gyroscopes is one of the more elegant pieces of pure science that you could ever contemplate. Its application in the iBOT and the Segway HT are about simulating human balance. In the building of the iBOT 3000 Mobility System, we inadvertently discovered the Segway HT. We realized that since our gyroscopes are more sensitive than the inner ear, that our computers are faster than typical reflexes, and that our motors are more powerful than the muscles in our legs, we could use these technological enhancements for the able-bodied too.

When did you know you wanted to make things that really helped people?

I feel that if I'm going to spend months and usually years and sometimes – in the case of the iBOT – decades on anything, it has to be worth it

for me. It has to be something that, if successful, makes a difference in a meaningful way. I just couldn't put time, energy, and passion into things that don't matter.

Tell me about the work you do with FIRST (For Inspiration and Recognition of Science and Technology).

As radically different as the Segway HT is from a scooter is how radically different FIRST is from a science fair. I founded FIRST on the premise that in a free society you get what you celebrate.

When the machine fits the mission, I believe that technologies improve the quality of our lives. The problem with this last mile – the niche distance between walking and driving – is that nobody until now had the right technology to apply to it.

The average kid can tell you the names of famous living athletes and Hollywood entertainers, but they'd be hard pressed to tell you the name of a single famous living scientist or mathematician or engineer. FIRST is a yearly science competition modeled on a sporting event. It creates demand among kids to excel at math and in science and builds incentive for them to get involved with the opportunities of the time.

What's next?

Oh boy, what's next! I got myself in trouble the last time I talked about something before it was done. But I can say that right now I continue to work on what I think are important pieces of technology that have medical applications and will improve people's lives. We're working on ways to do for water and electricity what the cell phone did for communications – most especially for people in the developing world, who will likely never have an electrical or water grid. We hope to help them leapfrog over the twentieth century utilities infrastructure right into the twenty-first century, so that they can be on par with the developed world.

Dean Kamen is the president of DEKA Research in Manchester, New Hampshire.

Moving together: The new mobility culture considers not only transit but also health, education, housing, waste, and social needs. No transportation system is an island; it must coordinate all shared systems for maximum effect.

We cannot talk about urban transport until we know what type of a city we want. How do we want to live? Do we want to create a city for humans or a city for automobiles? The important questions are not about engineering, but about ways to live.

– Enrique Peñalosa, former mayor of Bogotá, Colombia

Archetypal cities of the new mobility culture include Bremen (Germany), Bogotá (Colombia), and Curitiba (Brazil). Bremen's mobility strategy takes its inspiration from a mythical creature known as *"die eierlegende Wollmilchsau"* ("egglaying-woolmilksow"), which translates roughly as an all-in-one device that's suitable for everyone. The idea is to develop an intermodal system, involving public transport and car sharing. In Bogotá, just two years since the implementation of Transmilenio, the BRT (Bus Rapid Transit) system, there have been radical improvements in mobility and overall quality of life: decreases in travel time for users (32%), violent crime citywide (50%), traffic accidents (80%), number of fatalities caused by traffic accidents (30%), and noise pollution (30%); and an increase in time spent by mothers and fathers with their children (37%). With Curitiba, its population has more than tripled since 1965 (going from 500,000 to 1.6 million), and, because of inspired leadership by architect and former mayor Jaime Lerner, it continues to support its citizens across the board. Innovative bus systems carry 1.3 million people daily (up from 25,000 per day 25 years ago). The city also pays its homeless population to keep the parks clean, and runs an exchange system allowing people to swap garbage for food and transport vouchers.

2.11 THE AURA OF SUBWAYS, THE COST OF BUSES. Integrated mass transit system in Curitiba, Brazil.

Jaime Lerner
on public transport

Curitiba, Brazil, is considered one of the best examples of urban planning on the planet. When did you begin participating in the design of its master plan?
In the mid-1960s, I was part of a group of architects working for the city of Curitiba, advising the mayor at the time (Ivo Arzua Pereira) in every development phase of the Curitiba Preliminary Urban Plan. We later became the Instituto de Pesquisa e Planejamento Urbano de Curitiba (IPPUC), the Curitiba Research and Urban Planning Institute. Through IPPUC, I participated in the preparation of the master plan to guide the city's physical, economic, and cultural transformation, and was elected mayor of the city in 1971. I remained mayor for three terms (1971–75, 1979–83 and 1989–92).

How can a city be an instrument for change?
A city has to have the political will to change. A city needs a strategy that works with potentiality, not just needs. And a city needs solidarity, not as rhetoric but as a sincere understanding of the daily life of its citizens. With every problem there needs to be an equation of co-responsibility. When everyone understands what the consequences of certain attitudes are, they will more readily cooperate and help bring about change. A city needs to have a daily plan and daily processes that encourage constant learning.

This is why you have designed initiatives where the citizens are involved in such things as tree planting, recycling, and keeping the gardens clean?
Yes, involvement in all aspects of city life. When we started separating garbage in Curitiba, we looked first to the children. For six months, we taught every child the importance of separating organic from inorganic garbage. The children then taught their parents. Since 1989, we've had 70% voluntary participation in this initiative. When we had a fuel crisis about 20 years ago, even though we had a very good system of transport in place, everyone knew that they should rethink public transport. Curitiba has more private cars than any Brazilian city except Brasilia (500,000), yet 75% of commuters take the bus and Curitibanos spend only 10% of their income on transport. Why? Because they have a good alternative.

Jaime, tell me about the significance of the Plexiglas bus tubes you designed.
It was said for so many decades that a good system of transportation should be underground. But when you don't have the financial resources to build such infrastructure, it helps you to have more creativity. The tube, a less expensive option, gives the buses of Curitiba the same performance as a subway. We started to study this about 30 years ago and knew what was needed to create a good system of transport: it had to be fast, reliable, comfortable, and with good frequency. This meant not only putting buses in exclusive lanes like in many cities of the world, but also allowing for boarding on the same level and paying before getting on the bus. The tube supports both. In 1974, we moved 25,000 passengers per day with buses running in an exclusive lane. The system was improved regularly and now we are transporting more than two million passengers per day.

On surface, we can have better frequency and the connections are faster. Underground, you can travel faster, but it's technically impossible to have a frequency less than two minutes and the connections take longer; sometimes it takes 15 minutes or more to walk underground alone. I have nothing against subways, but the problem is that it's hard to have a complete network of underground systems. Even cities that have a few subway lines need an effective surface system. The future of mobility has to be considered in terms of integrated systems, where each piece – bikes, cars, taxis, subways, buses – never competes in the space of another.

What do the different colors of the buses signify?
The colors allow for easy reading of the system. The double-articulated buses are red. The feeder buses are yellow. The interborough buses are green. So you know by the color of the bus what kind of bus it is. If you want to build solutions for the future and have people working with you, every citizen has to understand the system very well. You have to have a commitment with simplicity. Every child should know the design of his or her own city. They should design the city even, because if you can design the city you can understand the city. If you understand the city, you will respect the city.

Even cities that have a few subway lines need an effective surface system. The future of mobility has to be considered in terms of integrated systems, where each piece – bikes, cars, taxis, subways, buses – never competes in the space of another.

What is the capacity of a double-articulated bus?
300 people. I used to joke about this though: it's 300 Brazilians or 270 Swedes! As soon as the system has a very good boarding process, you can transport 300 passengers every minute very easily, which is 18,000 passengers per hour in one direction. In thirty seconds, this is 36,000 people per hour, which is a subway statistic, which is what we do in Curitiba.

Do you believe this kind of transformation can occur in any city of the world?
I'm convinced that every city in the world, no matter the scale, no matter the financial resources, can have a significant change in less than two years. I can swear to you that this is possible and that you can make important changes. With public transportation, you can definitely make important changes. With environmental issues, you can make important changes. On the care of children, you can make important changes. It all depends on the city, but anything is possible. It's not a question of scale. Sometimes mayors use as an excuse that their cities are too big. But no, it's not a question of scale; it's a question of philosophy. Don't be afraid of the scale. Don't be afraid if you don't have enough financial resources. You can always build a good equation of co-responsibility.

Are other places in the world mimicking the Curitiba model?
It took 25 years until another city tried to do what we did in Curitiba, and that city is Bogotá, Colombia. Now, more than 83 cities in the world are doing it, including São Paolo. I was in Seoul, Korea, and the mayor there is removing highways in the city in an attempt to design a surface system and restore an old stream, a small river in the city that was very important to their history. Honolulu is trying to implement a surface system and I'm sure they will do it very efficiently. These are just two examples of many, many more.

This must give you a feeling of great satisfaction.
It does, for sure. I'm convinced that the most important thing to work on right now is the mobility system, which is not only a system of transport; it's the whole understanding of a city. The more we create an integration of functions, the better a city will become. We don't yet have a smart car. We have a smart bus, which is a good system of transport.

Cars that are good on design are not good on the engine. Cars advanced on the engine – hybrid systems – are not good on design. Small cars are not good on the road. Cars for the road are not good for the city. Bikes are another issue. We have to redesign the bike so that it opens up like an umbrella. If bikes were more portable we could integrate them with public transport, going from door to door. I'm working on this idea with a team here in Curitiba. We're exploring how far we can go with surface transportation. Lastly, why not develop a kangaroo system for cars? Rather than having two cars – one for the road and one for the city – we could have a car that feeds energy into a very, very small car.

I like to say, cars are our mechanical mothers-in-law. You have to have a good relationship with your mother-in-law, but you cannot allow her to conduct your life.

Jaime Lerner is an architect in Curitiba, Brazil.

2.12

2.13

City	Car	Bus	Differential
Belo Horizonte	23	16	7
Brasilia	45	27	18
Campinas	24	17	7
Curitiba	22	19	3
João Pessoa	26	18	8
Juiz de Fora	30	21	9
Porto Alegre	29	20	9
Recife	24	14	10
Rio de Janeiro	26	19	7
São Paulo	16	11	5

Average evening peak car and bus speeds in Brazilian cities. In Curitiba, where there's a negligible three-miles-per-hour differential between car and bus, public transit – with none of the hassles of parking – is, in fact, the better way.

2.13 ONE PLAN FITS ALL. Curitiba's street plan from 1943, when its population was only 140,000. By 1965, it had grown to 500,000 and tripled in size again by 2001 to 1.6 million.

2.14 GROWING STRONG. In 1974, 25,000 people per day took the bus. Today, more than two million riders per day move around Curitiba this way. Left to right: the growth of Curitiba's public transportation system, from 1974 to 2000.

2.12 and **2.15** INSPIRED LEADERSHIP. Jaime Lerner's plans and oversight made Curitiba a widely hailed model. Biarticulated buses whisk up to 300 passengers along their commutes at 90% of the speed of automobiles.

2.16 FAST. Opening a new chapter in railroad history, Shanghai began operating the world's first commercial Transrapid Maglev train in January 2003. Developed in Germany, the vehicle levitates over a 30-km stretch of double tracks, between Long Yang Road Station and Pudong International Airport Station – traveling at 430 km/h. Three vehicles, each with six sections, will carry 10 million passengers in 2005, 20 million by 2010, and 33 million by 2020.

2.17 FASTER. In Japan, the Yamanashi Maglev Test Line runs 42.8 kilometers between Sakaigawa and Akiyama. The three-car, manned Maglev vehicle MLX01 has been undergoing feasibility tests since 1996 and achieved a record speed of 581 km/h in December 2003.

2.18 THE MAGLEV EFFECT. The map of North America warps to show the effect of increased speed over distance. If a Maglev MLX01 were to run the 2,800 mile (4,505.2 kilometer) distance between New York City and Los Angeles, a 45-hour car ride would become a seven-hour Maglev ride, just about as fast as a nonstop flight.

2.16

2.17

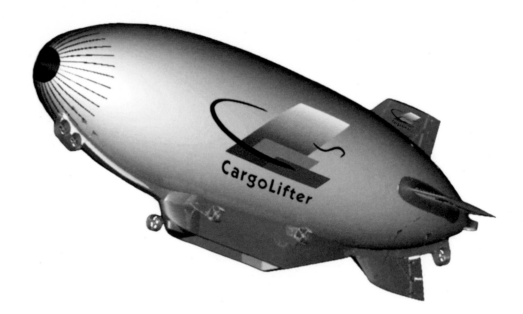

2.19 BEYOND THE END OF THE LINE. The CargoLifter will carry heavy loads (up to 176 tons) through the air to areas previously impossible to access. Its diameter is 71 yards. Its length is 284 yards. And its total weight is 463 tons.

2.20 BEYOND THE BOUNDS OF GRAVITY. NASA's Institute for Advanced Concepts (NIAC) supports the effort to build an elevator to space. The Institute for Scientific Research (ISR) estimates that, in fifteen years, a space elevator capable of lifting five-ton payloads every day en route to all Earth orbits, the Moon, Mars, Venus or the asteroids will be operational.

Off the grid: In our ongoing project to colonize the universe, the outer boundary has been the edge of infrastructure, the end of the road. Today we continue to push beyond that threshold, to boldly go.

If we have learned one thing from the history of invention and discovery, it is that, in the long run – and often in the short one – the most daring prophecies seem laughably conservative.
– Science-fiction author Arthur C. Clarke, *The Exploration of Space*, 1951

Moving off the grid is about moving beyond existing infrastructure – beyond the confines of roads, rail and other well-trod delivery routes.

Berlin, Germany–based CargoLifter has begun a project to design a helium-filled airship that functions as a flying crane. The CargoLifter 160 (CL 160), once implemented, will be able to transport up to 176 tons of cargo at a speed of 56 mph over a distance of up to 6,214 miles. The amazing thing about the CL 160 is that it remains in the air when the motor is switched off and it doesn't have to land for the cargo to be loaded and unloaded, because of its patented load exchange process.

The Institute for Scientific Research (ISR) in Fairmont, West Virginia, led by Dr. Bradley Edwards, hopes to make the space elevator a reality. The idea is to enable travel from Earth into space on a ribbon of carbon nanotubes – one end anchored to Earth and the other end out beyond geosynchronous orbit (35,800 km altitude). Electric lifts, clamped to the ribbon, will shuttle payloads to space at an unbelievable cost savings: the space elevator lift costs are predicted to be $100 per pound, where current launch costs are $10,000–$40,000 per pound.

2.21

2.22

2

2.23

2.21 and **2.22** EKIP AWAY. With the U.S. Naval Air Systems Command (NAVAIR) as a new partner, the Saratov aerospace engineers might finally get to see the global adoption of the EKIP. It uses a vortex-oscillating propulsion system that enables it to take off and land at steep trajectory (on land and water). The EKIP has a saucer-shaped body with short, fat wings, boasts air-cushion landing gear, and embeds its main power units and auxiliary engines inside, rather than out. It is constructed of lightweight composite materials and, most importantly, can run on natural gas and hydrogen.

2.23 NON-FLYING SAUCER.
Avro-Canada designed the Avrocar in the 1950s. The jet-powered saucer captured the attention of the U.S. Air Force, which took over the project in 1955. In the end, the project ceased because of design instability; the Avrocar's elevation capped out at a few feet.

2.24 IT'S A ... BIRD. The SiMiCon Rotor Craft is a hybrid Unmanned Aerial Vehicle out of Norway with the vertical takeoff and landing of a helicopter and the high-speed capacity of a fixed-wing aircraft (350–450 km/h). Its design is based on the arctic tern, a bird known to hover 30–40 feet above water before nose-diving and submerging itself to catch small fish.

Global movement: The success of the plane has brought problems to the skies and the atmosphere similar to those that the car has brought to the streets. With ever-increasing global air travel, sustainable solutions are a must.

Forecasts of air travel made by Airbus Industrie and Boeing, the two major manufacturers of commercial jet aircraft, indicate that global air travel alone will continue to grow at a rate of just under 5% a year for the next 20 years.

– Mobility 2001 report, World Business Council for Sustainable Development

Air travel is the fastest-growing means of global transportation. But it faces severe environmental and operational sustainability problems. It exceeds its share of energy use because the pollutants are released at high (and more absorptive) altitudes and shifting to non-carbon-based fuels is less feasible for air transportation than it is for automobiles.

Since 1980, however, 430 miles southeast of Moscow at the Saratov Aviation Plant, a saucer-shaped aerial vehicle named EKIP has been going through design iterations, with a mission to solve the problem of energy-inefficient air travel.

Perhaps the EKIP can be to the sky what the Segway is to the city. An off-airfield, amphibious, high-efficiency, environmentally safe, new type of flying vehicle, it has the potential to revolutionize aviation. Lots of experiments have come before it and after it – the Boeing Flapjack, the Avrocar, the SiMiCon Rotor Craft (SRC) – but, like the Segway, the EKIP stands out in the field. It promises to do away with long stretches of concrete runways, noise pollution, and carbon-based jet fuels. The EKIP was developed behind the Iron Curtain, then perestroika made it public. In 1995, the project was patented in Russia, Canada, Europe, and the United States. Test flights are slated for 2007 at Webster Field, near Patuxent River, Maryland.

We will bring energy to the entire world.

Extreme ultraviolet Imaging Telescope (EIT) image of the Sun.

We continue to design massive hydroelectric and petroleum projects with regional and even global economic, social, and environmental impact, building machines on a scale the world has never seen. At the same time, initiatives for sustainable energy — wind, geothermal, and especially solar — promise to fundamentally restructure the energy system itself, from a centrally based system to a distributed network of energy production and consumption.

ENERGY ECONOMIES

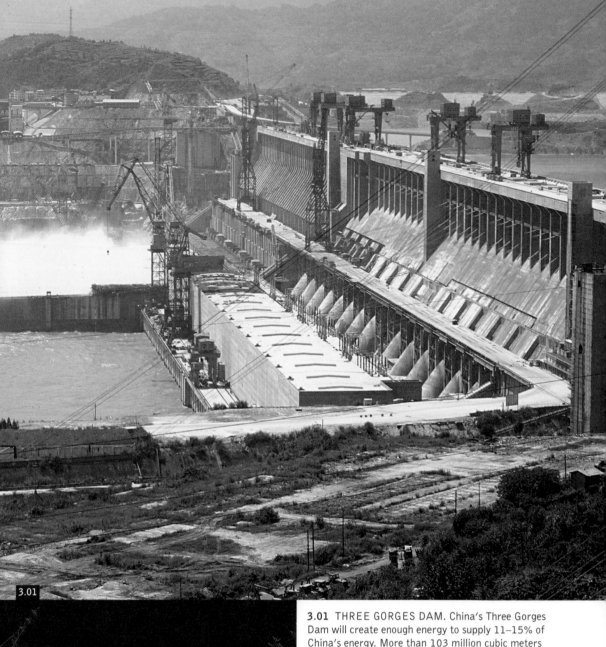

3.01

3.02

3.01 THREE GORGES DAM. China's Three Gorges Dam will create enough energy to supply 11–15% of China's energy. More than 103 million cubic meters of earth and rock were excavated to build the dam and another 28 million cubic meters of concrete will be used to complete it by 2009.

3.02 ONE GORGEOUS DAM. Completed in 1935, the Hoover Dam still provides power today to California (56%), Nevada (25%), and Arizona (19%). It took four years to build, cost $165 million, and reaches 221 meters high. Although the Hoover Dam measures a whopping 379 meters wide, the Three Gorges Dam will be five times this width.

The big projects: In the realm of energy, we are seeking out small, portable options for our decentralized future. Meanwhile, we persist in designing feats of wonder that will fundamentally reconstruct the natural world.

The problems of the world cannot possibly be solved by skeptics or cynics whose horizons are limited by the obvious realities. We need people who can dream of things that never were.
– U.S. President John F. Kennedy, speech to the Irish Parliament, Dublin, June 1963

The Itaipú Dam on the border of Brazil and Paraguay generates 12,600 megawatts of electricity, which makes it the largest hydroelectric power plant in operation in the world today. By 2009, this will no longer be the case. China's Three Gorges Dam, on the Yangtze River, will be the world's largest and highest dam, whose 26 turbines have an installed capacity to generate an estimated 18.3 gigawatts of energy. Long admired as a region of beauty, the Three Gorges – named after the towering limestone cliffs of the Qutang, Wu, and Xiling gorges – stretch 200 kilometers through four provinces in China's heartland and inspired even Mao Zedong to write poetry. Now the site of the largest hydroelectric project in the world, it dams the third longest river in the world, causing it to rise 600 feet high and drown ancient temples and tombs. This effort will also affect the lives of 1.3 million people in the end, making it the most significant resettlement program ever attempted. If the dam breaks, all hell will break loose: the polluted waters will flood all the major cities along the Yangtze riverfront. And although it's a question whether or not the human rights of the relocated people were considered in the transition, 15 million people downstream will be better off because of the newfound electrification. If the dam succeeds, it will save China from burning 50 million tons of coal per year.

3.03

3.03 CHILLY WATER IN, CITY WATER OUT.
Toronto-based Enwave and the City of Toronto collaborated to bring naturally occurring cold water (4° C water at 83 meters' depth) to the dense urban area, with the view to meeting downtown air-conditioning needs without relying on fossil fuels. The infrastructure, the world's largest lake-water cooling service, is a submerged 5-km pipeline in Lake Ontario designed by Griffin International Engineering. Compared to conventional air-conditioning methods, this system requires 75% less energy and eliminates 40,000 tons of carbon dioxide per year.

3.04 IT'S HIGH, MATE. EnviroMission is developing Solar Tower technology for the Australian outback. This project will be the world's tallest engineered structure. One 200-megawatt, 1-km-tall solar thermal renewable-energy power station will generate enough electricity for around 200,000 typical Australian households, and will cut approximately 700,000 tons of greenhouse gases from entering the environment every year. *Time* magazine voted it one of the best inventions of 2002.

CHILLED WATER H4 IN

CITY WATER H3 OUT

3.04

3.05 LOOK OUT, LOCH NESS. Researchers from The Robert Gordon University (RGU) in Aberdeen, Scotland, are using a prefabricated tidal current device called SNAIL, which can be cheaply installed in shallow and deep water, to generate energy. Inventor and world authority on tidal energy, Professor Ian Bryden says that tidal currents offer a substantial and predictable source of renewable energy. The Scottish resource alone, if developed effectively, could provide enough electricity to support a population of 15 million. Once commercially available, the SNAIL is expected to exploit the massive energy potential that exists in the Pentland Firth and smaller tidal sources in sea lochs off the west coast of Scotland and the inter-island channels of Orkney and Shetland.

3.05

ENE73

3.06 FASTEST*. The Z pulsed power accelerator at Sandia National Laboratory in New Mexico began operating in its present form in September 1996. Along with predecessor machines, it was originally designed to simulate certain aspects of the explosion of nuclear weapons. Z is the world's most powerful and efficient laboratory X-ray source. Its most powerful pulse was more than 270 trillion watts, more than 80 times the combined output of all the Earth's utility plants. Time-exposure photography shows the electrical arcs produced across the water-air interface in the accelerator tank, where 20 million amps pass through a target the size of a spool of thread, reaching velocities that would fly a plane from Los Angeles to New York in a second. These flashovers are much like strokes of lightning but are actually run-off of the main, unseen current flow. The goal of a bigger model would be to capture neutrons in an electrical power plant and create cheap energy for the entire world. *See page 62.

3.06

3.07 THE BIRDS AND THE BEES. Green technologies power British Petroleum's Hornchurch Connect site. It not only runs entirely on renewables, but it also features sustainable water management and landscaping to attract local wildlife, such as dragonflies and insect-feeding birds. Reed-bed technology is used to filter and clean waste water (the bacteria living on the roots breaks down oil contaminants), rainwater is collected and recycled, and the wildflower turf under the wind farm provides a habitat for bumblebees.

3.07

3.08

3.08 FILLING UP ON WATER. According to Stuart Energy, the hydrogen economy is already here. Its method of water electrolysis eliminates fossil fuels from the production cycle, enabling countries to use indigenous resources (renewable electricity and water) to produce clean, zero-emission hydrogen. The vision for distributed, on-site generation of hydrogen through water electrolysis began when Alexander T. Stuart and his son, Alexander K. (Sandy) Stuart, formed the company in 1948. Today, the Stuart Energy vision continues.

3.09 FOSSIL FUEL–FREE. The first Shell-branded hydrogen fuel station to operate anywhere in the world opened in Reykjavik, Iceland, in April 2003.

3.09

Clean green power: Initiatives for sustainable energy promise to fundamentally restructure the energy industry itself. To stay in the game, the oil industry is evolving away from fossil fuels, towards renewables. The most massive change will happen here.

In our generation, we will see something happen that has never happened before in our history: the peaking of the most important single energy source.

– Richard E. Smalley, Nobel Prize–winning chemist

In the 1960s, the late geophysicist Dr. M. King Hubbert famously foretold the end of American oil production. Based on his research, he came to the conclusion that oil production would peak within a year or two of 1970. He was widely abused for this notion. But it turns out, in fact, that he was right. Current projections indicate that global oil production, too, ought to peak well before our population stabilizes – somewhere around mid-twenty-first century, with fertility rates dropping as they are. Natural gas and nuclear are the current forerunners to replace oil. But, in close analysis, neither proves adequate. Natural gas will not be able to meet the world's energy demand much beyond 2020, and a future of nuclear breeder reactors isn't exactly an attractive scenario. So major players like Shell, British Petroleum, and Stuart Energy have been working toward alternatives to oil: wind, solar, and hydrogen fuel by way of water electrolysis.

Among numerous renewable initiatives, Shell is developing *Noordzeewind*, a wind park off the coast of Wales, which will consist of 36 turbines (99 megawatts total) and provide electricity for 110,000 households. British Petroleum's Hornchurch Connect site (near London) runs entirely on renewable energy, generating up to half of its own power from solar panels and wind turbines. Stuart Energy builds hydrogen fueling stations around the world, enabling the production, storage, and delivery of on-site emission-free fuel.

3.10

3.11

3.10 and **3.11** A MARKET OF TWO BILLION PEOPLE.
The Turbo Stove is an innovative design by Tapio Niemi
of Helsinki, Finland. It's lightweight, inexpensive, and
can be assembled in fifteen minutes without tools. It
burns fuel efficiently and operates on bioenergy: all vari-
eties of solid fuel (peanut shells, corn-husks, straw, and
animal dung) in addition to wood. Niemi designed the
Turbo Stove over a ten-year period from the experience
he gained while part of the Sudan-Finland Forestry
Program. He says, "Locally produced recyclable biomass
remains the best fuel source for cooking for millions of
the world's rural population. It is environmentally sustainable and alleviates dependency on expensive and
environmentally harmful fossil fuels, which accumulate
CO_2 content in the atmosphere."

3.14 THE FIRE DOWN BELOW. Iceland has tapped
into the fact that more than 99% of the Earth's volume
is more than 1000° C hot. Using this knowledge and
Iceland's abundant geothermal energy potential, more
than 90% of homes in the country are heated from the
ground up.

3.12 PALM SPRINGS BLOWS HARD. This windmill farm along Interstate 10, in California's San Gorgonio Pass, generates enough electricity (600–900 MW) to power the entire city of Palm Springs and the surrounding area. The nearby San Jacinto mountains, the third-highest mountain range in California, provide a wind tunnel along the highway, providing steady gushes of hard-blowing winds – enough to power the windmills 300 days a year. Wind energy is a commercially available renewable energy source. The top wind producers in the world are Germany, the United States, Spain, Denmark, and India.

3.12

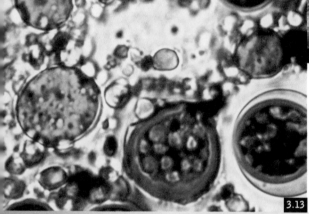

3.13

3.13 WEE GAS STATIONS. Researchers at the U.S. Department of Energy's National Renewable Energy Laboratory (NREL) are conducting experiments on microalgae, microscopic aquatic plants that can produce up to 60% of their body weight in lipids (oils). From these lipids, which are the golden color seen in this image, a diesel-like fuel can be derived.

3.14

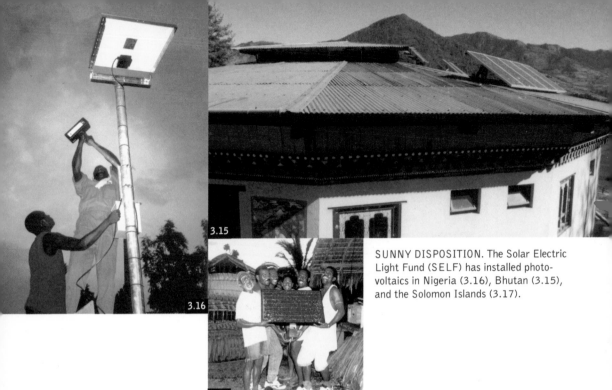

SUNNY DISPOSITION. The Solar Electric Light Fund (SELF) has installed photo-voltaics in Nigeria (3.16), Bhutan (3.15), and the Solomon Islands (3.17).

3.18 COPYCATS. Researchers at ASU's Center for the Study of Early Events in Photosynthesis study photosynthesizing bacteria with this question in mind: "Can we create a better solar cell based on this?" Ana Moore's synthetic solar cells mimic the essential primary processes in natural photosynthesis.

Solar power: All energy is solar energy, stored in different forms. Every two minutes the sun gives the earth more energy than is used annually world-wide. It is the only renewable resource with the capacity to provide all the energy we need on a global level.

The human species now has a major challenge: to make solar energy cheaper than coal. And there's no question in my mind that with human ingenuity and five thousand times more solar income from the sun than we need to operate, we will solve the energy problem.
– William McDonough, architect and one of the leaders of "The Next Industrial Revolution"

English-born American physicist and mathematician Freeman Dyson pointed out in his book *The Sun, The Genome, and the Internet* that solar energy is most abundant where it's needed most: in the countryside rather than in cities, and in the tropical countries, where most of the population lives, rather than in temperate zones.

On the global scale, the Solar Electric Light Fund (SELF) provides technical and financial assistance for solar energy and wireless communication systems in the developing world. SELF has launched what it calls "solar rural electrification programs and enterprises" in all corners of the globe. These programs have brought photo-voltaic systems – and often electricity for the very first time – to remote areas of China, India, Sri Lanka, Nepal, Vietnam, Indonesia, Brazil, Tanzania, Uganda, South Africa, and the Solomon Islands.

On the atomic scale, the Arizona State University (ASU) Center for the Study of Early Events in Photosynthesis researches ways of capturing light and transforming energy. The center is comprised of a multidisciplinary group of scientists, one of whom is synthetic organic chemist Ana Moore, who looks to purple bacteria – which can harness the sun's energy through photosynthesis – to better understand these processes and how we might use them as a model for developing improved solar panels one day. The results of this research illustrate the advantages of designing functional nanoscale devices based on biological paradigms.

Robert Freling
on rural
electrification

How is SELF helping to power small villages all over the developing world?
We use photovoltaics as the primary technology – solar cells that convert sunlight directly into electricity without burning any fossil fuels. This technology has been around since the 1950s, when NASA developed it to power satellites in space. Over the last few decades, there has been a steady advance in the efficiency with which these cells convert sunlight to electricity and in improved manufacturing processes, which help to lower their production costs. It has reached a point now where we can visit rural villages all over the world, places that have little hope of ever being connected to a conventional electric grid any time soon, and install panels directly on the homes, schools and clinics and generate clean solar electricity.

Two billion people still live in the dark. How can solar energy help to bring electricity to the entire world population?
Photovoltaics offer many benefits to the developing world. Because the panels can be quickly installed in very remote areas, you have an opportunity to bring immediate relief at the household level. This is what we did as a primary mission during our first decade; we focused on household lights using typically 50-watt systems, which generate enough power to run three or four lights, a radio and a couple of appliances in the home. As a result, we noticed a rapid increase in the quality of life. No longer were these people breathing in toxic

kerosene fumes, children were studying and reading at night, and entire families were engaged in productive activities during the evening hours.

How does SELF financially assist the families in acquiring these solar home systems?
Even though we're a nonprofit organization, we did not believe that giving the systems away outright was a smart thing to do. On the other hand, asking families to pay $400 or more in cash for a solar home system is just prohibitively expensive. So we used microcredit and other innovative forms of financing that would allow these families in developing countries to pay for and take ownership of these solar electric systems over a period of time. We proved through a series of pilot projects in about eleven countries that if you can provide the access to credit, rural credit, many of these families – who are already spending five dollars or in some cases ten dollars a month on candles, kerosene, and small dry cell batteries to power radios – were able and willing to pay for solar electricity.

Why combine solar power with wireless communications technology?
Together, they allow us to bring not just light and power to remote communities around the world but also access to voice and data connectivity. We've done this in South Africa and, more recently, in a remote part of the rain forest in the Amazon. In collaboration with a conservation group called the Amazon Association, we installed solar power systems to run lights, water pumping systems, refrigerators, computers, and a satellite dish to deliver high-speed broadband Internet access to a group of Caboclo Indians. It has really made a difference in their lives.

Explain the idea of "the center is everywhere."
Good concept. In today's world, our consciousness is dominated by this notion of center-periphery. If you live in New York you have access to information, entertainment, people, restaurants, and more. If you're out in the middle of the Gobi Desert,

3.19 SELF brings power to the Amazon.

you're often at a severe disadvantage in terms of your access to information and entertainment. The world that I envision, which is being made more and more possible through distributed power and wireless communications technology, is a world where no matter how isolated you are, you have the same opportunities and access to information as the people in New York do. It's inexcusable to me that two billion people – a third of humanity – are living without light. We have to find a way of more equitably distributing information, and using science and technology to promote social justice.

How does it make you feel to bring light to villages around the world?

It's exciting to travel to these remote villages and spend time in them and witness firsthand the transformation that can occur with just a little bit of help from SELF. The local communities drive our projects. These people embrace what we do and are well organized and looking to take greater control over their own lives. Being able to generate power for themselves is a wonderful first step and often gives them increased hope and confidence in improving other aspects of their lives. Just seeing a light come on for the first time can be very powerful. I've seen it in village after village where people celebrate this event. And it goes beyond this – now we're using the technology to pump purified water and store vaccines, which is still a problem in terms of bringing immunization programs to remote areas where there's no electricity. Vaccines have to be stored between 0° and 8° C, and if the cold chain is broken the vaccines are lost. The World Health Organization and others have demonstrated that solar-powered vaccine refrigerators can be a very viable option for storing vaccines and medicines.

What happens on cloudy days?

The panels collect solar energy and convert it into electricity, which is then stored in batteries. On cloudy days or at night, you're drawing from battery power. There are other applications, such as water pumping, where you don't need batteries. In this case you can simply pump water to a tank and this tank becomes your storage medium. But with anything electronic – lights, computers, radios, electronic equipment, satellite dishes – batteries are used to store the power from the sun.

Are photovoltaics most cost-efficient off the grid?

For rural communities living off the grid, it is often the least expensive solution for providing electric power. In many cases it will cost over $20,000 per mile to extend the grid, which is why so many villages around the world are not connected; it just doesn't make economic sense for the utility company to extend the grid to remote areas where the population density is low and the use of electric power is limited. It makes much more economic sense to go in and install these distributed stand-alone systems that can provide exactly as much power as you need. Another advantage of photovoltaics is their modular and scalable nature. You can start off with a small system and, as the needs of a household or a community grow, you can add incrementally to generate more power.

It's inexcusable to me that two billion people – a third of humanity – are living without light. We have to find a way of more equitably distributing information, and using science and technology to promote social justice.

SELF currently runs projects in China, India, Sri Lanka, Nepal, Vietnam, Indonesia, Brazil, Tanzania, Uganda, South Africa, and the Solomon Islands. How far off are we from narrowing the gap between the haves and the have-nots?

It's going to take a lot of effort to really bring power to two billion people, but I do think it's an achievable goal within the next decade or so. There has to be political will by all the countries in the world to solve this problem. It's not beyond our reach.

Robert Freling is the president of the Solar Electric Light Fund (SELF), a nonprofit organization that promotes solar electricity and energy self-sufficiency in developing countries.

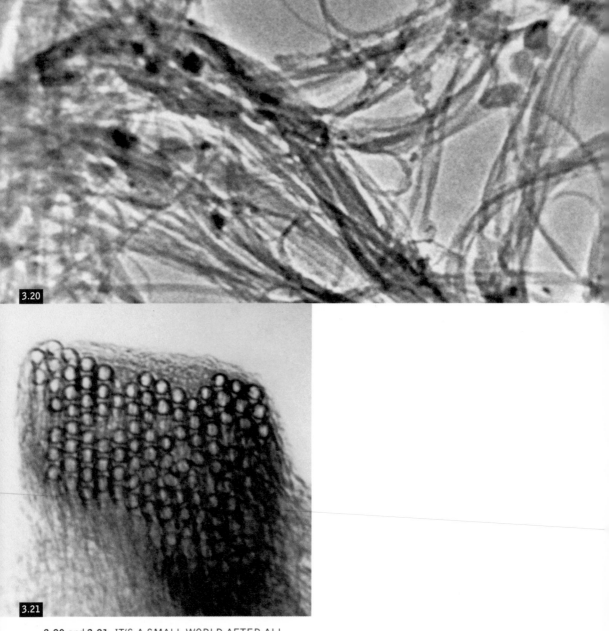

3.20 and **3.21** IT'S A SMALL WORLD AFTER ALL.
Transmission Electron Microscope (TEM) images of the
doubling of single-walled carbon nanotubes produced by
The Smalley Group at Rice University in Houston, Texas.

The worldwide grid: The future of our planet depends on our redesigning the current power system, which relies on large-scale, centralized entities. We need to produce energy locally and distribute it globally.

Electrical-energy integration of the night and day regions of the Earth will bring all the capacity into use at all times, thus overnight doubling the generating capacity of humanity because it will integrate all the most extreme night and day peaks and valleys. From the Bering Straits, Europe and Africa will be integrated westwardly through the U.S.S.R., and China, Southeast Asia; India will become network integrated southwardly through the U.S.S.R. Central and South America will be integrated southwardly through Canada, the U.S.A., and Mexico.

– Buckminster Fuller

The worldwide energy grid was the highest priority of Buckminster Fuller's World Game – an immersive experience in advanced design science that took place in the summer of 1971. The agenda was to explore "expeditious ways of employing the World's resources so efficiently and omni-considerately as to be able to provide a higher standard of living for all of humanity – higher than has heretofore been experienced by any humans – and on a continually sustainable basis for all generations to come, while enabling all of humanity to enjoy the whole planet Earth without any individual profiting at the expense of another and without interference with one another, while also rediverting the valuable chemistries known as pollution to effective uses elsewhere, conserving the wild resources and antiquities." The World Game challenged doomsday predictions and assumed we could strategically design solutions, with existing resources, to accommodate the entire human population.

Nobel laureate Richard E. Smalley of Rice University has this very same dream. The worldwide grid he envisions is made of single-walled carbon nanotubes, which are, not surprisingly, also known as "buckytubes."

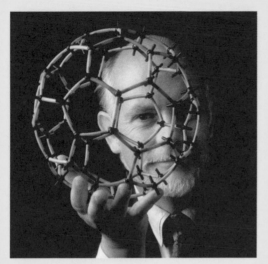

Rick Smalley
on our energy
challenge

Rick, tell me about C_{60}.
C_{60} is also known as the "buckyball." Imagine a soccer ball with its seams arranged in a pattern of pentagons and hexagons. If you sit and count the vertices, you'll find that there are precisely 60 of them – just like the carbon-60 molecule. Sixty is a remarkable number. It turns out that it is the largest number of objects you can arrange around the surface of a sphere and have each be identical to every other one by a simple rotation. Sixty also has more integral factors than any other number and any range of numbers, if 60 is inside that range or some multiple of 60. It is this factorability of 60 that gave the Babylonians reason to use it as the basis of their number system. And it's the reason that around the world we divide the hour into 60 minutes and the minute into 60 seconds.

What sort of special genius does the carbon atom have wired into it?
Carbon has a genius for forming a one-atom-thick membrane, the sheet of which graphite is made. In this graphite sheet – called graphine – each carbon atom is connected to just three others, very much like the vertices in the hexagonal mesh of chicken wire. In graphite, carbon stays defiantly 2-D in a 3-D world. Carbon is the key connective entity in organic molecules and there's a way of binding its atoms together that not only gives you the versatility of organic chemistry but also magnificent electrical conductivity in the form of buckytubes, which are the best conductors of

electricity of any molecule that we've ever discovered. The buckytube is for electrons what a single-node optic fiber is for photons. It's a single-node waveguide for the transmission of electrons.

I understand your research group at Rice University has the motto "If it ain't tubes, we don't do it."
That's right! And we've recently formed The Carbon Nanotechnology Laboratory, whose tag line is "We're going to make buckytubes be all that they can be."

And what is that?
Not only do buckytubes have incredible electrical conductivity, but we believe that these tubes will almost certainly have the strongest, highest tensile strength of any fiber ever made in the universe. The thermal conductivity down these tubes has been measured to be at least 50% higher than diamond, which previously had the highest-known thermal conductivity. So the electrical properties are unmatched, and the mechanical and thermal properties are unmatched. Our challenge is to make these incredible properties of the individual nanometer-wide tubes become manifest on the macroscopic scale, in real-world applications.

What types of tools do you work with at the nano scale?
We produce nanotubes by way of chemical tricks. We don't actually go in with magic fingers and pick up individual atoms and build these tubes one atom at a time! We now have six different ways of making buckytubes – some of which are being commercialized right now. And we continue to look for new ones. What we'd like to get – and I'm confident we will – is an industrial-scale method that will allow us to make a particular type of buckytube with perfection, in hundreds of millions of tons a year around the planet at low cost.

3.22 A direct-band gap semiconducting nanotube undergoing a chemical trick.

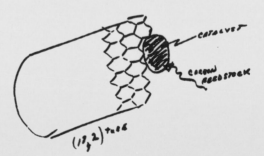

Where are we now in terms of the application of nanotech to real-world problems?
That really depends on how you define nanotechnology. In my definition, we are already very far along that road. I view nanotechnology simply as a technology whose power is determined entirely by our ability to manipulate atoms at the nanometer scale and build structures. But the overall structure need not be nanometer in size; it could be something big enough to hold in the palm of your hand.

I can imagine that there is, in fact, at least one good clean answer to how we can provide the energy we'll need for about ten billion people on the planet by 2050. I suspect that over the next ten years we will be able to get a major research program of the magnitude of Apollo to make this come to pass.

Seen this way, modern engineered polymers fit the definition. Even nylon would have been a nanotechnology in its day. And pretty much every drug that you take is a nanometer-scale molecular entity which, if you changed any atom in it, wouldn't work the same way. These are all nanotechnologies. The new stuff with buckyballs and buckytubes is within its first decade and will take another decade or so before it has a huge impact.

I know you're passionate about the world's energy challenge. How are you feeling about it these days?
I'm feeling pretty good because I can imagine that there is, in fact, at least one good clean answer to how we can provide the energy we'll need for about ten billion people on the planet by 2050. I suspect that over the next ten years we will be able to get a major research program of the magnitude of Apollo to make this come to pass. There are probably on the order of ten miracles that need to happen: stunning breakthroughs in the physical sciences and engineering that will enable the world to run on a new kind of oil – an energy technology for the twenty-first century that will be the basis of prosperity, as oil was in the last century. I believe that this new energy technology, if you had to describe it with a single word, would not be hydrogen. It would be electricity as the connective tissue that brings us all together. We're headed toward a world powered primarily by solar energy, wind, renewable biomass, and likely quite a bit of nuclear. But, increasingly, the lion's share of the world's energy will come from the sun directly: photons that hit our houses, giant solar farms in the great deserts of this world, and solar energy beamed down to earth in the form of microwaves from solar collectors in orbit around the earth – or perhaps even on the moon.

When we imagine a global energy grid, what is the role of nanotechnology?
Take the interconnected electrical energy grid we have now – in North America, for example – and add a local storage facility for electrical power at every house and every business. These would function like uninterruptible power supplies, providing 12 to 24 hours of buffer from whatever happens on the grid. This means that your peak load is taken care of from your own storage and you buy energy off the grid when it's cheap, at night. Primary power would come to the grid from existing and new power plants. To this scenario, we add efficient long-distance transmission of electrical power, which would certainly not be carried along aluminum or copper wires. It will be a much more efficient wire. One way or another, nanotechnology's got to find the answer.

Tell me about the magnificent opportunity with our energy challenge right now.
I believe that in the energy debate we're having around the world – and very intensely right now in North America – there is a high ground that's currently not occupied by any of the principal political parties. It's a high ground because it is not selfish. It's pro-development and pro-environment. Taking on our energy challenge now will do more to address the rich/poor divide than any other single thing. I can think of no greater mission than to use this wondrous new nanotechnology that's being developed to solve humanity's need for clean energy.

Richard E. Smalley is a Nobel Prize–winning chemist and professor at Rice University in Houston, Texas.

We will build a global mind.

The most profound impact of information technology has been to transfer
the potential of the scientific method –
the ever-expanding accumulation of
knowledge – to the cultural sphere.
Internet protocols allowed us to link any
two computers, enabling an explosive
global network of networks. Emerging
grid protocols for distributed computing
allow us to link everything else – data-
bases, simulation and visualization tools,
and the unused computing power of
machines – generating a worldwide
cultural accumulation beyond imagination, available to anyone, anywhere.

INFORMATION ECONOMIES

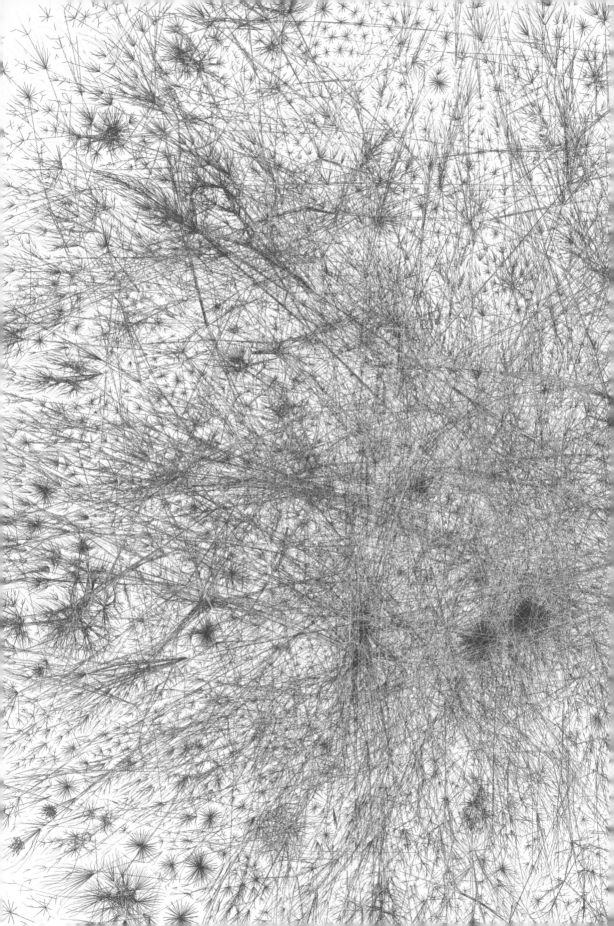

4.01 THE MOST DANGEROUS THING ON
EARTH (AND THE MOST BEAUTIFUL).
This graph (made in November 2003
by The Opte Project) makes visual our
radical new possibility for knowledge.
The project's founder, Barrett Lyon,
chief technology officer for California-
based DigiDefense International, wrote
a software program with the capacity
to map the entire Internet in a single
day, by a single computer. The collected
information provides an analysis of
wasted Internet Protocol (IP) space,
maps IP distribution, and detects the
results of natural disasters, weather,
and war. The latest research shows
that the known Internet is growing by
more than 10 million new, static pages
each day. The effect is the global accumulation of knowledge over time and
the ever-increasing value of search
engines like Google.

4.02 and **4.03** THE PENGUIN AND THE GNU. Tux, the penguin, is the mascot for the Linux movement. Linus Torvalds, founder and lead programmer of Linux says, "Really, I'm not out to destroy Microsoft. That will just be a completely unintentional side effect." The gnu is the mascot for the Free Software Foundation (FSF), whose motto is "GNU = Gnu is Not Unix." Richard Stallman, founder and lead programmer of the FSF, admits, "If someone were pointing a gun at me and demanded that I write non-free software or else, then I would do it. But when his back was turned, I'd escape. And I'd probably put in bugs."

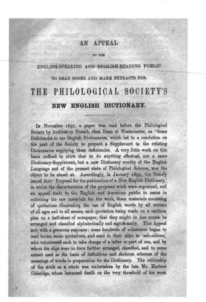

4.04 and **4.05** THE OED AND CHUCK D. The open-source and free software efforts today are modern-day examples of the sort of collaboration that went into the making of the Oxford English Dictionary (OED). The first page of the "Appeal for Readers" was distributed across the English-speaking world. Now, for those who like to burn, Napster salutes you. And so does legendary hip-hop artist Chuck D, Public Enemy frontman. Debating the Napster controversy with Metallica's Lars Ulrich on *The Charlie Rose Show* on PBS in May 2000, Chuck D said, "This is industry versus the people, and the people got technology on their side and we gotta adapt." Napster, once the rebel peer-to-peer (P2P) online music sharing service, is now a partner with Microsoft and Yahoo, hoping to bridge corporate profitability with the convenience of downloading MP3s.

Learning to share: To imagine that any one closed group could solve the complex problems we face today is folly. The free and open software movements promise to overcome our territorial attitudes and take advantage of our collective potential.

Information doesn't want anything, but people want it to be free so that they can trust it. Hidden information always makes you wonder who's hiding it and why.

– Esther Dyson, chairman of EDventure Holdings

In his book *Tomorrow Now*, Bruce Sterling talks about information now seeming like everything to everybody. He writes, "Information is commerce, media, politics, science, art, education, military power, a good, a service, a dessert topping, a floor wax, porn…" We're clearly soaking in it. Its abstract pervasiveness, though, should not make it any less important – in terms of scrutinizing its contents – than anything else. We ought to create enough of a critical distance from information that we can ask questions about the operating forces that bring it into being.

In a free society, all code should be liberated. We want to know the programming that sits behind the fancy package. But we don't want it to get in the way of what we want it to do for us. We want it to be transparent. When it fails, we know it. And if we don't have access to the code, we have no way of knowing how to fix it. When it fails, and the code is available to us – as with the open source and free software movements – then we can see where the bugs are and fix the problem. It really comes down to having access to the information – not that the majority of us will know what to technically do with code. But we need to begin caring about the significance of code and the importance of its effects. After all, as Mitch Kapor, designer of the hugely successful Lotus 1-2-3 spreadsheet program and creator of the Electronic Frontier Foundation (EFF) – a public interest group working for the civil liberties of people in information technology – says, "Architecture is politics." If we remain ignorant to code, we inadvertently remain ignorant to politics.

Lawrence Lessig
on free culture

How are coders themselves increasingly becoming lawmakers?

In implementing and choosing the architectures that will define cyberspace, you're implementing and choosing certain architectures to enable or disable values. So you're making political choices. What's troubling is when these political choices are made by entities that aren't responsible publicly; we then begin to worry about the extent to which this kind of private lawmaking defeats public values.

How is commerce changing the character of the Internet?

The intended consequence of "cookies" is to deposit little markers on your hard drive so that the website "remembers" you and what you want to buy, which makes it easy to shop online. The unintended consequences include the fact that it's now much easier to track people as they move around the Internet, and to target advertising or gather information from people. What are typically considered invisible markers are actually indelible markers. And as a result, it's now very easy to monitor and chase all online transactions. Those in business are not paid to think about privacy and personal liberty. Businesses do what they're paid to do: find ways to make it so they can sell stuff on the Internet. We need to think about the consequences of their techniques, and if those consequences corrode values that are important to us. If so, then we have to find ways to resist this.

What are the roots of intellectual copyright law?

People have an understandable view that the idea of copyright has been around for 200 years and that it has never changed. And so, when you see this explosion of peer-to-peer file sharing – which is said to violate copyright laws – most people's natural response is to say, "Let's stop the theft." But in fact, there's a long tradition to consider. There was once a powerful group in England called the Congor. They were a monopolist group that restricted the spread of knowledge by keeping prices of books high. Then along came the Statute of Anne, which was designed to promote education and learning by limiting copyright to 14 years. Its effect was to basically tell the Congor that their government-granted monopolies would be over, and they would have to compete in the marketplace if they wanted to continue to prosper. As a result of its implementation, for the first time in English history, the works of Shakespeare, for example, were no longer under the control of monopoly publishers. Works became free and the tradition of free culture was really born.

What do the open source and free software movements of today have in common with the Oxford English Dictionary?

The OED was the most explicitly open source publishing project that we had before the production of the Free Software Foundation's new Linux operating system. It got born when there was an announcement in the newspapers around England that said, "We want to put together a people's dictionary" – an empirical dictionary. It wouldn't try to tell people how they were supposed to speak; it would try to figure out how people were using the English language and catalog that so there would be a common reference point when people were trying to understand what the language was or how it had changed. So they sent out a request for volunteers and literally got thousands of volunteers across England to begin to produce little note cards of the usage of the English language. These notes were sent to an editor who went through them all and collated the work into what eventually became the OED. It's an example of creativity that I think is not limited to software dictionaries but is common in many areas of creative life.

So your interest in copyright is not exclusive to cyberculture?

The war in cyberspace over copyright has increased the regulations of copyright, but the burden of those increased regulations is not just felt in cyberspace. Unfortunately, creators everywhere feel it. This is the aim of the free culture movement: to try to remove the unproductive, burdensome restrictions so that we can get back to a world where it's easy for creators to create. Especially when technologies like the Internet enable a wide range of creators to create.

Was it Walt Disney who changed everything?

Walt Disney's greatest work built on the public domain. Walt Disney took the stories of the Brothers Grimm and retold them in a warm and fuzzy way, and captured the imagination of many generations of Americans. He was free to take those stories and retell them in the way that he did because the Grimm fairy tales had passed into the public domain. This was Walt Disney's technique – and it's been the technique of the Disney Corporation all the way to the present. Because Disney has been so successful in extending the terms of copyright, nobody can do to Walt Disney what Walt Disney did to the Brothers Grimm. Nobody can build on top of Disney's work in the way that Disney built on top of other peoples' work. And that change in the basic bargain of copyright is what I think has been most destructive to the way in which free culture has evolved. Free culture has always depended upon the Walt Disneys of the world having the freedom to build without seeking permission upon our past. That freedom has now been removed by lobbyists, who convince Congress that a better way to have a culture is to require that you first get permission from corporate owners.

What do you mean when you say, "A time is marked by the ideas that are taken for granted"?

If you really want to understand a culture, don't look to the things people argue about but, instead, try to understand the things they take for granted. I believe there's an important role to be played by copyright, in creating incentives for a culture to be developed, but we have, almost subconsciously, adopted an extreme vision of copyright protection. The result is that we are destroying the very opportunities that copyright was originally intended to enable.

What are your thoughts on Copy Left?

Copy Left is the creation of Richard Stallman and the Free Software Foundation. They use the rules of copyright to not limit the spread of culture but guarantee that creative work is freely available for people to build upon.

How is Microsoft responding to this?

Microsoft, of course, is a company that was built entirely on proprietary software, not free software, and so it's threatened by the extraordinary growth and employment of free software. Both by governments and commercial entities, who find, for example, the GNU/Linux operating system to be just as good and in many ways better – and certainly cheaper – than the Windows operating system. Microsoft has been very keen to drive governments away from the General Public

> Because Disney has been so successful in extending the terms of copyright, nobody can do to Walt Disney what Walt Disney did to the Brothers Grimm.

License by, at various times, suggesting that it would destroy the country's software industry and claiming that it was an unfair business practice for the government to prefer open source or free software versus proprietary software.

Your thoughts on the future?

I hope the future is one where people increasingly embrace the idea of making content available on freer terms than content is available now. I am the chairman of something called the Creative Commons, which enables people to mark their content with licences that signal freedoms associated with the content, just like the Free Software Foundation tried to do with software. My hope is that we get a much wider range of adopters to this model, so that the extremism of All Rights Reserved, which was Hollywood's vision – versus No Rights Reserved, which is the kind of anarchist's vision and is no longer what defines the debate – is replaced by something more moderate, something that enables artists to build and share content, but also compensates them for their creativity.

Lawrence Lessig is a professor of law at Stanford University in Palo Alto, California.

4.06

4.06 CERN'S NEW PET PROJECT.
The Large Hadron Collider (LHC) particle
accelerator at CERN – the world's largest
particle physics laboratory, based in Geneva,
Switzerland – will become operational in
2007. The experiments that use the LHC
will have unprecedented computational
requirements and generate 12-14 petabytes
of data each year. The site will unite the
supercomputing power of sites across
Europe, America, and Asia.

4.07 SUPERCOMPUTER OF SUPER-
COMPUTERS. Launched by the National
Science Foundation (NSF) in August 2001,
the TeraGrid (right) is the world's largest,
most comprehensive distributed infrastructure for open scientific research. It includes
20 teraflops (20 trillion floating point
operations per second) of computing power
distributed across five sites. Each facility
is capable of managing and storing nearly
1 petabyte (1,024 terabytes) of data, high-
resolution visualization environments, and
toolkits for grid computing.

4.07

The scientific project: Information now travels along vast interconnected networks, and so all problems in all realms are shared. We need to learn a lesson from science and do what it's been doing all along: distribute problem solving.

Computing and data grids are emerging as the infrastructure for twenty-first century science because they are providing a common way of managing distributed computing, data, instruments, and human resources.

– William E. Johnston, chief architect, NASA Information Power Grid (IPG)

Grid projects are designed to be distributed information systems whose computing goals can range from complex forms of interaction in the medical community (bio-informatics) – where researchers work toward the development of cures for diseases and viruses – to the analysis of seismic activity, as with the NEESgrid (Network for Earthquake Engineering Simulation). What distinguishes a grid from other distributed computing systems – or the Internet, for that matter – is its concern with integrating distributed resources using standard protocols and interfaces to provide capabilities that would not otherwise be available.

Developed by computer scientists Ian Foster and Carl Kesselman, the Globus Alliance is a consortium of institutions working on a software system called the Globus Toolkit, which is focused on defining, developing, and assimilating an open source implementation of some of these core grid protocols.

The grid is exciting because it's an evolutionary step beyond today's Internet. It builds on the same technology that underlies email and Web browsers, but it extends that technology to allow us not just to access information as the Web does, or to send messages as the email system does, but also to tie together computing systems that may be geographically dispersed.

Ian Foster
on grid computing

How is grid computing today similar to the Internet in the early nineties?
In the early nineties, the Internet made a transition from being something of utility, and really only known to people in academia, to something that was of broad industrial relevance with the emergence of the Mosaic web browser and Netscape. Grid computing is making a similar transition at the moment, from academia to industry. Just in the last two years or so we've seen major corporations, like IBM, Sun, and HP, deploying grid products.

What is the history of distributed computing?
Grid computing is about the large-scale integration of computing systems to enable new classes of applications to provide on-demand access to computing and information. And it's certainly not a new idea. Back in 1969, when the very first node of the (then) ARPANET, which became the Internet, was deployed at UCLA in Los Angeles, a press release went out touting the wonderful things that were going to happen once the Internet was ubiquitous. It was quite an ambitious and visionary view of things, given that they only had one node at that point! What's different now is that we have the quasi-ubiquitous Internet networks suddenly getting fast enough that we can connect our computers and people and information sources in ways that were not possible before. Also, the software has evolved to the point where we can start thinking about linking

distributed computing systems into something really interesting.

When did our computer processing capacity and data storage capacity reach the point where we could even consider implementing distributed computing utilities or grids?
One important part of the evolution towards grid computing is, in addition to the deployment of the Internet, the fact that our home computers are now as powerful as yesterday's supercomputers. As we all know, the power of our computers continues to double every eighteen months or so. The laptop that I use for most of my work nowadays, for example, is faster than the supercomputers that were deployed at the U.S. national centers just ten years ago. This process is an ongoing one and the same observation will probably be true ten years from now. But the transition to a system that enables participation in a grid as a true peer is something that's just happened in the last five years or so.

What was it like back in the day when you and your colleague Dr. Kesselman envisioned designing software that could juggle and link all the computing resources across sites and deliver them on demand?
Our thinking on these topics, of course, has evolved over the years, but I think the vision has stayed fairly consistent. We started on this in the mid-nineties when the high-speed networks were starting to be deployed. We had a strong interest in understanding how computers, and distributed systems for that matter, could be used to enhance the process of scientific discovery. And we realized that there was a need for a new class of technology that would allow people to build distributed computing systems more easily than was then the case. So we put together a small team, coined the name Globus, and we started working on some basic software with scientific teams who were eager at the time to try new technologies. It has gradually scaled up from a few partners to an increasing number of large scientific projects and, more recently, industrial groups.

What is the difference between grid computing and distributed computing schemes such as SETI@Home?
SETI@Home is a simple and very effective example of grid computing. It is one of the larger examples of a number of systems that sort of behave in similar ways, whose goal is to harness idle computers in order to perform a large number of computational tasks that are essentially

identical but different in terms of the data that is being processed. In the case of SETI@Home, the problem is that of searching for extraterrestrial intelligence by processing data from the Arecibo radio telescope. Every home computer that signs up receives periodically a piece of radio telescope data and processes it using signal processing algorithms that look for particular signatures that correspond to a signal. With SETI@Home, the goal is indeed large-scale integration of computing systems, but the actual communication patterns that are going on are pretty simple. Grid technology involves far more complex couplings of systems at different locations, and often much larger volumes of data and more challenging computational tasks.

One important part of the evolution towards grid computing is, in addition to the deployment of the Internet, the fact that our home computers are now as powerful as yesterday's supercomputers.

What is the TeraGrid?

The TeraGrid is one of the most ambitious scientific grid projects underway. This is a U.S. project funded by the National Science Foundation. It involves the construction of very large computer systems at four sites across the U.S. – two in Illinois and two in California – and the linking of those systems by a very high-speed network, one that runs at 40 gigabytes per second, which is a million times faster than a typical home modem. These four sites and a number of others are integrated into a grid system so that a scientist with access can sign on and obtain on-demand access to any one of the systems in this grid infrastructure.

So it's the linking of supercomputers?

Yes. This is one important application of grid computing but certainly not the only one. There's an interesting project called Grid2003 that links more medium-scale computer clusters at thirty sites or so across the U.S., and across to computers in Europe as well. Grid2003 is intended for large-scale data analysis in the physics community.

How is NASA tapping into the grid?

NASA is an early adopter. They created a project some years ago called the NASA Information Power Grid (IPG), which aims to link together some of the major NASA laboratories. Some of their goals involve the design of aerospace vehicles; they're interested in allowing distributed teams to formulate and then run numerical simulations of these systems. One of the challenges they face is the number of different components involved: there could be an airframe model, an airflow model, and the combustion model for the engine, each developed by different groups. Grid computing links them together to build a numerical simulation of the ensemble.

How does the Globus Toolkit software change the way earthquake engineers work?

Earthquake engineers used to have to travel great distances to facilities to construct and run experiments on shake tables, very large structures on which you put model buildings and simulate the effects of earthquakes, and the results were typically fairly closely held. As a solution, the National Science Foundation in the U.S. has funded the Network for Earthquake Engineering Simulation (NEESgrid), which links together the earthquake engineering facilities in the U.S. with their remote user community, and with various data archives and simulation computers, to form what's called a "collaboratory," which is another word for grid, if you like.

What grid developments are you most excited about?

I remain delighted by the range of applications in the sciences, as with the earthquake engineers, the astronomers, and more recently the biologists. I think there are tremendous opportunities to accelerate work that will uncover the secret of life at a very fundamental level. It's also wonderful seeing what's going on in industry. In the next few years, I believe we'll realize this vision of a universal computing facility.

Ian Foster is Associate Director in the Mathematics and Computer Science Division at Argonne National Laboratory and the Arthur Holly Compton professor of computer science at the University of Chicago.

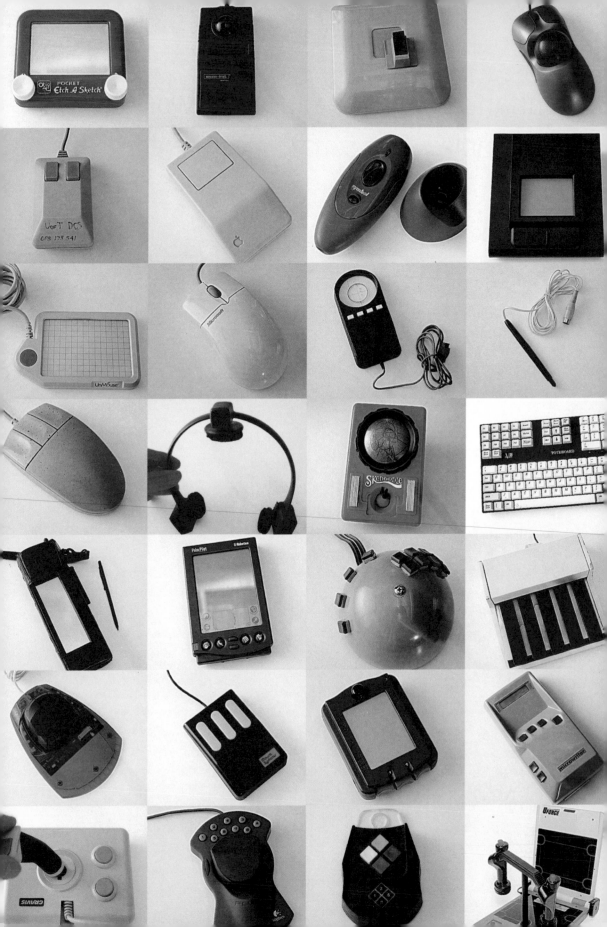

The interface: We need technological enhancements because the world is complex. But sometimes the complexities we encounter are a result of the technologies themselves. Good design augments human possibility and reduces complexity.

When we have problems interacting with technologies, it's a direct result of our not having asked the right questions in the design process. To be effective, we must shift our focus from the techno-centric to the human-centric.

– Bill Buxton, research scientist and principal, Buxton Design, Toronto, Canada

We function by making the most of the skills we've acquired from a lifetime of living. Motor-sensory skills help us perceive and move about in the world. Cognitive skills help us make sense of things. Social skills facilitate our participation in a community. Together, these skills make us human and are as essential as they are valuable. A well-designed tool honors this.

Every child in school today can learn to do long division because the Indians (via the Phoenicians) brought us the decimal point. The chronograph eliminated the need for extensive computation on the high seas to determine location. The microscope took us to a visual realm that was otherwise too small to see. The telescope let us view things that were too big and too far. The movie camera granted us the unfolding of visual events over time. Think of the delight in experiencing finely crafted musical instruments, paintbrushes, mountain bikes, cars. Now think about computers. They too have incredible potential to augment human capacity. But to do it well, our expertise about humans must at least match our knowledge of engineering. The human must be balanced with the technological or else invention becomes, as Henry David Thoreau said, "an improved means to an unimproved end." The challenge is as great as the potential.

4.08 A BOY'S TOYS. Assorted input devices, from the personal collection of Bill Buxton.

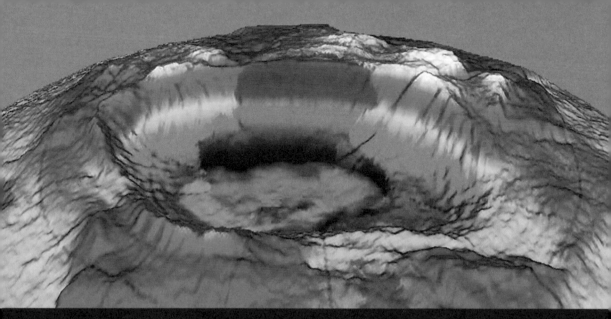

4.09 UH-OH. Using data derived from the Total Ozone Mappping Spectrometer (TOMS) instrument, this 3-D image of the hole in the ozone layer high above Antarctica was rendered at NASA's Goddard Space Flight Center in 1988.

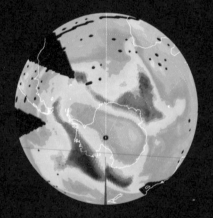

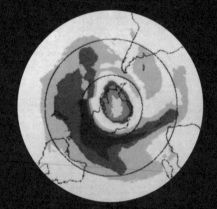

4.10 Dr. Joseph Farman discovered ozone depletion while NASA scientists were still calibrating their instruments. This first image of the ozone hole in 1978 did not cause alarm.

4.11 The first published ozone image captured the public's attention in 1983.

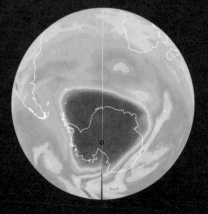

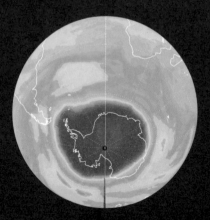

4.12 An early published ozone image mobilized the scientific community in 1985.

4.13 The key image that prompted the first global agreement to reduce usage of chloroflourocarbons.

Global portraits: The Montreal Protocol was spurred by the striking image of a hole in the ozone layer over Antarctica. It stands as a perfect example of how the international community can galvanize to tackle a global issue.

Perhaps the single most successful international agreement to date has been the Montreal Protocol.

– Kofi Annan, Secretary General of the United Nations

The discovery of the Antarctic ozone hole in the early 1980s by Dr. Joseph Farman helped bring about a breakthrough: the international agreement (The United Nations Montreal Protocol on Substances that Deplete the Ozone Layer) to reduce the global production of ozone-depleting substances. The images later compiled from data gathered by NASA's Nimbus-7 satellite solidified Farman's findings and delivered a poignant message to the general public that a critical link existed between ozone loss and chlorofluorocarbons (CFCs), manmade chemicals commonly used at that time in aerosol sprays and as refrigerants. It was because of the image that we became aware of the significant impact our actions have on the global environment. In conjunction with the Montreal Protocol, much research into the causes of ozone depletion and the biological effects of increased ultraviolet radiation (UV) exposure has resulted, in addition to collaboration by the scientific and industrial communities to find safe alternatives to CFCs.

Farman recently told BBC reporters that "there's progress in the sense that ozone-killing substances have generally stopped being released, but you have to realize that once they are in the air it takes a long time before they go away. So, no, things haven't got better, but the steps have been taken that will ensure that eventually they will disappear."

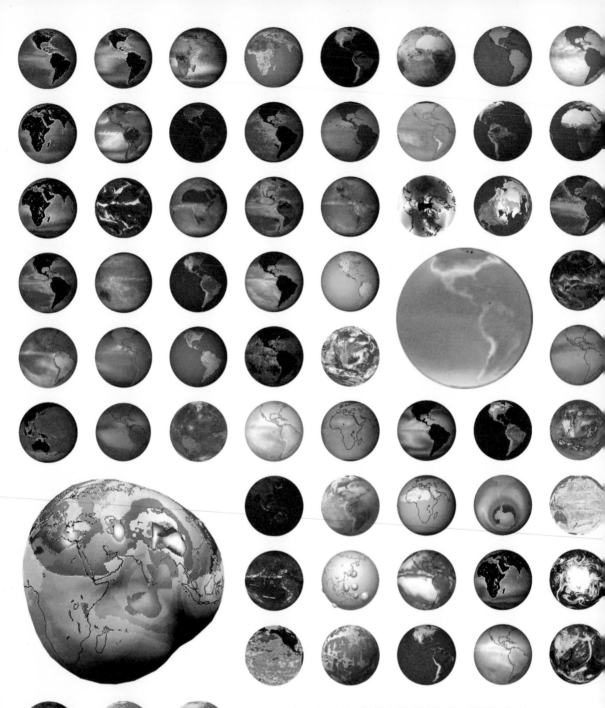

4.14 and **4.15** THE MANY FACES OF EARTH. NASA's Earth Observatory satellites and Japan's Earth Simulator (ES) capture complex data that is processed and rendered into myriad global portraits, each one telling a story about the Earth's patterns (radiation, oceans, weather, etc.). The ES Center is the size of four tennis courts. It's a 40-teraflop collaborative climate-modeling project by the National Space Development Agency (NASDA) of Japan, the Japan Atomic Energy Research Institute (JAERI), and the Japan Marine Science and Technology Center. Director-general Tetsuya Sato of the ES says, "With Earth Simulator, we are now able to understand the Earth holistically, with all factors entangling together at once, from microprocesses of clouds or snow to macroprocesses of atmospheric circulation, just the way Earth is."

4.15

Stewart Brand
on the long view

What was Bucky Fuller's reaction to your button campaign that asked, "Why haven't we seen an image of the whole earth yet?"
It was all because of LSD, see. I took some lysergic acid diethylamide on an otherwise boring afternoon and came to the notion that seeing an image of the Earth from space would change a lot of things. So, on next to no budget, I printed up buttons and posters and sold them on street corners at the University of California, Berkeley. I went to Stanford and back east to Columbia, Harvard, and MIT. I also mailed the materials to various people: Marshall McLuhan, Buckminster Fuller, senators, members of the U.S. and Soviet space programs. Out of everyone, I only heard back from Bucky Fuller, who wrote, "Dear boy, it's a charming notion but you must realize you can never see more than half the earth from any particular point in space."

I was amused, and then met him a few months later at a seminar at Esalen Institute in Big Sur, California. I sat across from his lunch table and pushed the button over to him, asking him what he thought about it. He said, "Oh yes, I wrote to that guy." I said, "I'm the guy. So what do you think? What kind of difference do you think it will make when we actually get photographs of the earth from space?" There was this slow, lovely silence. Then he said, "Dear boy, how can I help you?"

Why was this image so powerful?
It was motivating for a lot of people, I think, because it gave the sense that Earth's an island, surrounded by a lot of inhospitable space. And it's so graphic, this little blue, white, green, and brown jewel-like icon amongst a quite featureless black vacuum. Islands know about limitations. Bucky led me to this notion. He said people still think the earth is flat because they act as if its resources are infinite. But that photograph showed otherwise. Unless and until we find other flourishing planets, this is all we've got and we've got to make it work. There's no backup.

What would the iconic image be now?
I grew up with the image of the mushroom cloud, which was the first image seen as potential planetary Armageddon – one great big nuclear exchange, and there we would all be. We cowered in the shadow of that for 20, 25 years. It was thoroughly supplanted two years later by the image of Earth from space, and I have a feeling there's still a lot of changes to ring on that. I'm a little sorry that Al Gore's idea of putting up a satellite whose job it was to keep that photograph absolutely daily fresh hasn't come to realization because it's the sort of thing that would make it a little more here and now than the still photograph. As far as a new icon, the Long Now Foundation is trying to add one with a clock you can visit in the limestone cliffs of eastern Nevada and look at pictures of, which very plausibly would go on for 10,000 years.

Is this a strategic response to our bumper sticker culture?
Yes. When things are moving faster in a civilization, one of the things you can do to keep balance is look out for the slow things being tended to well. Attention tends to move toward things that move quickly, you see. We pay attention to the daily paper, to the next election or to the next financial report. That's great. You're supposed to do that. But if it takes up 100% of your attention, you will lose things like training the children and having decent universities and tending to the preservation of culture. The Clock of the Long Now will be a peephole of predictability through a deeply unpredictable series of events that will come at us in the future.

What do you think, Stewart? Will we see the end of war?
[Harvard archaeologist] Steven LeBlanc's book *Constant Battles* talks about our routine, organized conflict, driven by there not being enough to eat. Whenever we exceed carrying capacity,

we fight over scraps. It's the absolute norm for humanity. Occasionally, we would bump the carrying capacity up by inventing agriculture or invading a continent with new tools but very quickly we'd rise back up to carrying capacity and then get back to the same old dilemma: do you starve or steal? The obvious answer is stealing, and that usually involves fighting. And on you go. This relates to what [Harvard entomologist] E.O. Wilson has written about in a beautiful book called *The Future of Life*. He predicts we'll bump up against carrying-capacity issues very soon, sometimes locally, sometimes in very large regions. Even now you see it in Rwanda, with its dense

The Clock of the Long Now will be a peephole of predictability through a deeply unpredictable series of events that will come at us in the future.

population, degraded natural environment, pillaging, raiding, stealing, killing. LeBlanc does say at the end, "Look, for the last three or four hundred years warfare has become more organized, but the actual lethality has gone down drastically. It used to be that 25% of all young men would die in a war – and you'd get that until quite recently in Papua, New Guinea, and various places where the old forms of warfare hold. But one advantage of state warfare is that it, like agriculture, gets industrialized. On the one hand, it's terrible that more civilians get killed. On the other side, fewer people are actually dying as a result of combat. But that trend could reverse. And we're seeing potential for that.

I know the Global Business Network does some work with the Pentagon. What about the possibility of a long peace?
I love working with the Pentagon because they're the only entity I know that is completely eager to think in half-century terms. And there are several reasons for that. They're not a commercial entity, so they're not worried about the next quarter. They're not a democratic entity, so they're not worried about the next election. There really is a socialist economy in the military, and the people that you encounter at the senior levels are extremely bright. They've come up in a very tough meritocratic pyramid, and are trained throughout their lives to think globally.

What sort of scenario planning do you do?
One of the scenarios that developed in the course of our work with the Pentagon was what we refer to as a rogue superpower. We were looking at the various threats from rogue states and one of us said, "Let's see. What if you combined a lone superpower? What about rogue states? What if they're one and the same?" The answer is a rogue superpower! So we looked at this at great length and, lo and behold, in 2001 we received a call from a friend in the Pentagon. He said, "I think we've gotten to the rogue superpower scenario."

What were your thoughts on 9/11?
We were thinking that it was sort of right on schedule. It was horrifying for a lot of people who had been working both in the Clinton Administration and in Congress on the terrorism environment because we were saying for some time, "Look, the U.S. is not invulnerable in this." So a lot of us just groaned because we had already thought about it.

It's hard to stay optimistic sometimes.
The balance I find most pleasant to live with, and also useful in the world, is to be personally optimistic and globally pessimistic – in the sense of not becoming cynical but becoming focused on how things can go wrong and what we can do to fix them. By and large, I think there are plenty of reasons for optimism. Things have been getting better. So you can build on that. I would not want to live in even Rome at its height. Television's much better than it was in the Roman days!

Do you believe the current technologies are moving us in the direction of greater equity?
In aggregate, yes. [Editor-at-Large for *Wired*] Kevin Kelly likes to say that we take three steps forward and two steps back, but the net is that we've taken a step forward. The way Bill Clinton puts it is, "We're moving along and a lot of times we stumble and fall and back up and all that stuff. But as long as we're stumbling in the right direction, we're probably doing all right."

Multidisciplinary thinker Stewart Brand is the founder of the Whole Earth Catalog, *a cofounder of the Global Business Network, and a trustee of the Santa Fe Institute.*

We will make visible the as yet invisible.

Subatomic particle tracks in a bubble chamber at the Fermi National Accelerator Laboratory in Batavia, Illinois.

Our relationship with the image began through our natural aperture and its capacity to convert energy into meaning. As writer and designer Sanford Kwinter has pointed out, the human nervous system evolved in an environment where noticing change – the slightest difference in the surrounding environment – could mean the difference between life and death. So it is not surprising that our most developed cultural forms are practices of the visual.

But we could not and would not stop there. So much of life occurs beyond our natural visual range. Through scientific tools and methods, we have extended our visual bandwidth to colonize the full range of the

electromagnetic spectrum, from radio waves to gamma waves. Now life in all its glorious complexity, from the dynamic division of cells to the vastness of the entire universe in all of its pulsing vibrancy, has been rendered accessible to our visual capacity.

The scientific image evolves as a product of extraordinary new possibilities for exploration, far beyond the range of human vision. Meanwhile, mass adoption of the means for making and sharing images in the cultural realm continues to explode exponentially.

As cost approaches zero and access to image production and circulation becomes universal, new solutions begin to emerge. Our insatiable embrace of the image knows no bounds.

Time made visible: this Hubble Ultra Deep Field image shows galaxies as they were thirteen billion years ago.

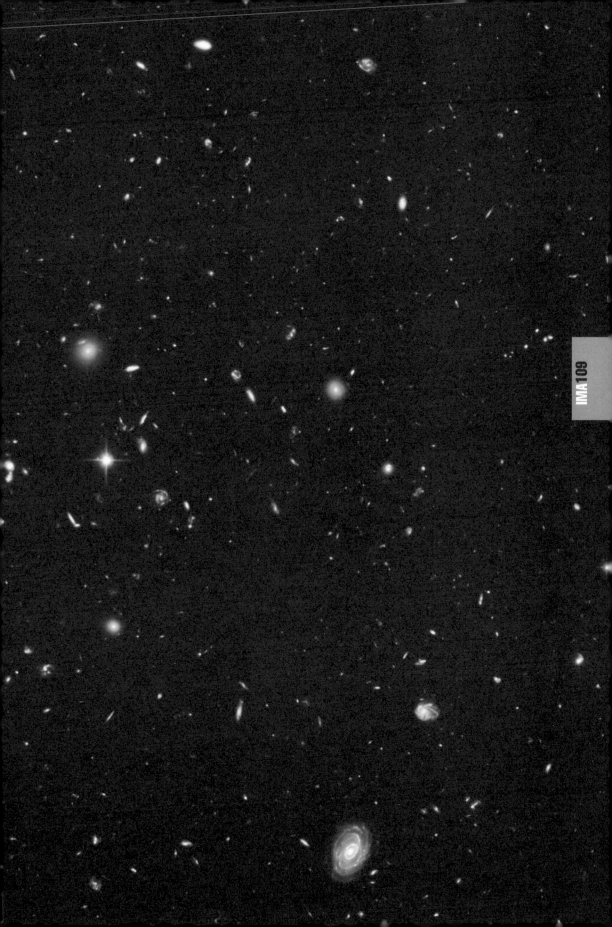

IMA109

David Malin
on astronomical
imaging

**How has your experience as a chemist –
working with optical and electron microscopy
and X-ray diffraction – influenced your work as
an astronomical photographer?**

When I started astronomy twenty-odd years ago,
I was very quickly involved with preparing photographic plates for use of the telescope; that's how
the data was recorded in those days, using photography. The things I learned and taught myself
as a chemist and X-ray diffraction expert were so
useful when I applied them in astronomy because
I was detecting radiation and trying to extract
from it the maximum amount of information. The
underlying principles are the same, but the technologies – telescopes versus microscopes – are
entirely different. The specialized hypersensitizing
techniques that I used in astronomy – the techniques that make the plates much more sensitive
to faint light – are quite complex; they involve
some knowledge of chemistry, and I had that
already, so it all came together beautifully.

**What is the electromagnetic spectrum (EMS)
and how does it relate to human vision?**

The electromagnetic spectrum is the full range of
radiation that permeates our universe and
extends from the very short wavelengths – X-rays
and Gamma rays – that we normally can't see
from the surface of the Earth, but we can generate from the surface of the Earth artificially –
right the way through to the longest, radio waves.
In the middle of this range of spectrum is visible

light; it's only a tiny part of the spectrum. And
that light illuminates our everyday life and is
divided up into the various colors that we see as
the rainbow. Human eyes are sensitive to just a
tiny part of the electromagnetic spectrum.

**How does the EMS factor into our daily lives as
human beings?**

If we go to the longer wavelengths, beyond
infrared, we enter the realm of microwaves – and
clearly they're an important ingredient in everybody's dinner! There are these practical uses.
But when we're able to detect [the wavelengths],
coming from outer space, they also tell us about
the nature of the objects that have emitted them.
Not so much microwaves, but slightly longer
wavelengths, centimeter wavelengths, which tell
us about the nature of the astronomical object
that we're looking at. Beyond these, radio waves
are the kind of wavelengths that we use for broad-
casting communication. They are received from
outer space, too, and tell us about the nature of
the stars and galaxies and the physical conditions
inside them. Depending on which wavelengths we
look at, we learn different aspects of the nature of
the natural world beyond the surface of the earth.
Shorter wavelengths of light, ultraviolet, are quite
an important diagnostic for many astronomical
objects. And shorter still, X-rays are a very important way of detecting highly energetic astronomical objects, but this has to be done from satellites
in earth orbit; it can't be done from the ground
because, rather surprisingly, X-rays don't pene-
trate the atmosphere very well.

**How do we extend that narrow band of vision
to include the invisible sea of radiation beyond
visible light?**

For radio waves, we use radio telescopes. These
are quite familiar to us as these large dishes –
hundreds of meters in diameter – usually steerable
arrays that we can point anywhere in space to
collect these rather feeble photons from radio
waves, and form them into streams of data that
we can analyze. At the shorter wavelengths,
we use microscopes and X-ray and electronic
diffraction equipment to understand the nature
of matter. So it's really a case of using special
equipment that's sensitive to these radiations to
be able to detect them and translate them into
ways in which our senses can understand them.

**What have been the most significant break-
throughs in imaging over the past few hundred
years?**

The inventions of the telescope and microscope,

which were around the same time, were extremely important because they extended the range of human vision, both to the stars and to the smallest things we can see. Their discovery led the basis of much of modern science because we suddenly became aware of a previously invisible world. The simple act of looking through a telescope at the stars and planets triggered a renaissance of thought that changed modern life, absolutely. Probably the next most important thing was the application of photography to scientific imaging, and that happened a hundred and fifty years ago or so; it was a slow process. Photography was discovered in 1839. But by 1880 it was being used in astronomy, and again it displaced the eye from the end of the telescope, because photography was much more sensitive than the human eye, and we could see much farther by using photographic plates. Photographic plates were also a detector of light for microscopes, which detected [light] and recorded it in a way that the eye never could. Photography was instrumental in the discovery and exploitation of X-rays and was deeply involved in the discovery of radioactivity. The most recent development that is revolutionizing the way in which we look at the world is the charged couple device, the solid state CCD, which we find in digital cameras and more.

What sort of impressive imaging devices are at work today at both the very small scale and the grandest of scales?
In the realm of the small, there are detectors that are used on the synchrotrons, which are devices that accelerate tiny particles to enormous velocities and crash them into targets; the atomic particles of those targets are scattered around in a very distinctive way. The scattering patterns, observed by special detectors (once known as bubble chambers), tell us about the nature of the particles themselves. At the other end of the scale, the Hubble Space Telescope has been making beautiful pictures of distant galaxies with distinctive arcs around them. These arcs tell us that those foreground galaxies have gravitational fields and that they are bending light from more distant galaxies – which means that the galaxies themselves are acting as huge cosmic lenses, creating arc-like, distorted images of much more distant galaxies that we could not see otherwise. These arcs represent the edge of our adventure at the moment in discovering the dimension of things in the natural world.

Is there anything beyond the EMS?
Radiation permeates the solid stuff that we call

5.00 David Malin's cluster of galaxies in Hydra.

the world around us and tells us about it in all sorts of ways. The electromagnetic spectrum is so enormously huge: it covers a range of 10^{26} in wavelength. I just can't imagine anything beyond it.

How can gravity aid in our quest to image the universe?
Its essential role is driving the universe. The stars, for instance, are held together by gravity, which compresses the core of stars so that they are dense and hot. This in turn drives the nuclear reaction within them, making the stars shine. But imaging gravity itself is an entirely different thing. Gravity waves are thought to be so enormously long and photons of gravity so weak that they cannot be imaged. But there are proposals to build telescopes to detect gravity waves, which will be generated when massive bodies such as black holes collide at enormous distances in the universe.

What do you feel is the driver behind our ever-expanding ability to make visible the invisible?
Curiosity. It's human nature to try and understand the world around us. We used to see through our telescopes a range of mysterious fuzzy objects and stars and all sorts of things in space. Now we know that these are galaxies and clusters of stars. None of this knowledge is of any commercial use, but it enriches our intellectual lives. It gives us a scientific framework within which we can discuss all kinds of philosophical ideas about the nature of the universe in which we live.

David Malin is an Adjunct Professor of Scientific Photography at Royal Melbourne Institute of Technology in Australia. He is the author of numerous books and articles and the scientific consultant to Heaven & Earth (Phaidon Press).

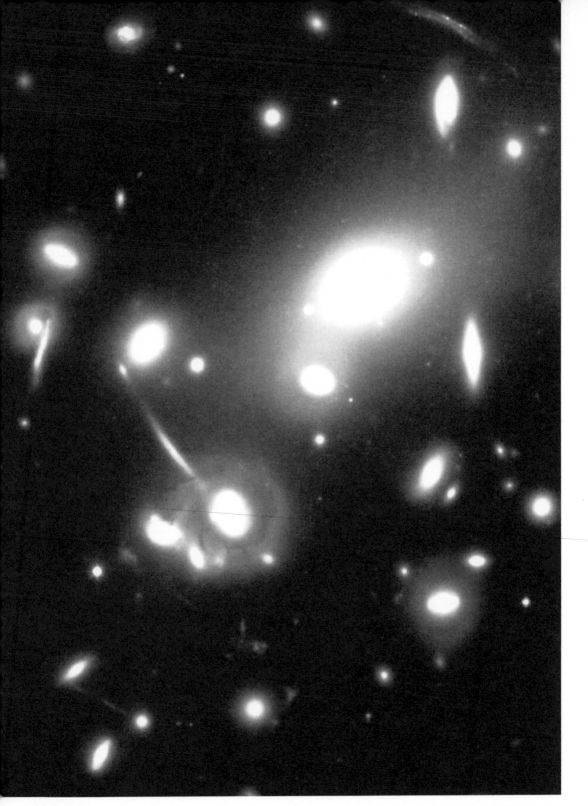

5.01 Galaxy cluster Abell 2218 functions as a two million-light-year-wide cosmic lens that bends light from galaxies over 13 billion light-years away.

5.02 The National Oceanic and Atmospheric Administration (NOAA) digitized radar image of Hurricane Hugo, near peak intensity.

5.03 Radio image generated by the Very Large Array telescope in New Mexico of Cassiopeia A, a remnant of a supernova explosion that occurred in our galaxy over 300 years ago.

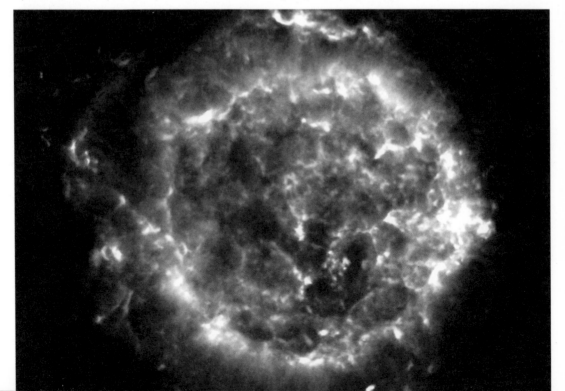

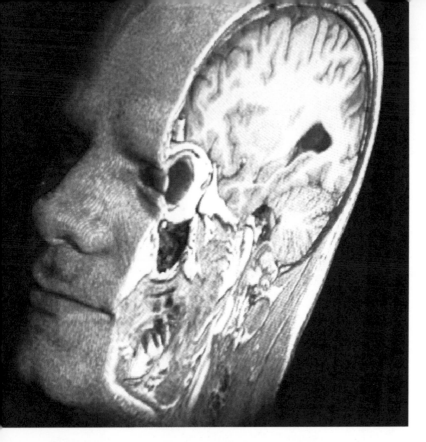

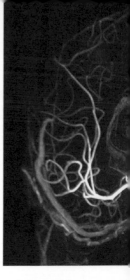

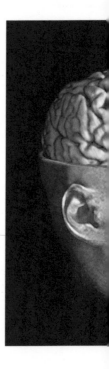

5.04 Magnetic Resonance Imaging (MRI) of the human head.

5.05 Model of the inner organs with more than 650 anatomical constituents derived from the Visible Human data at the Institute of Medical Informatics, University Hospital Hamburg-Eppendorf, Germany.

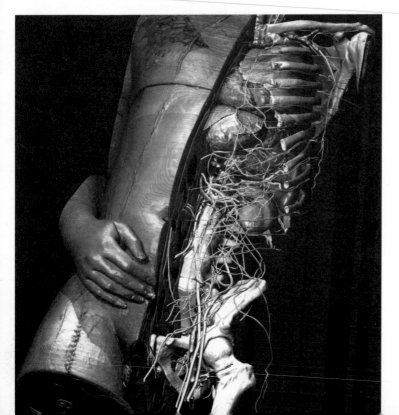

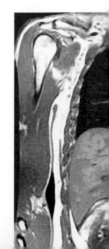

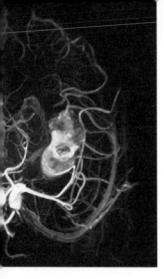

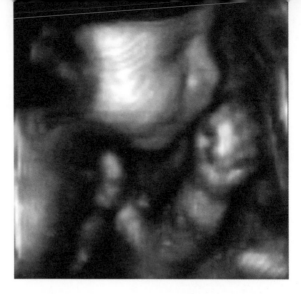

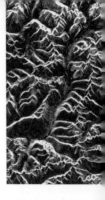

5.09 Radar image of south-east Tibet.

Satellite image of Kazakhstan (above).

5.06 Three-dimensional ultrasound image of baby in womb, produced beyond the range of human hearing.

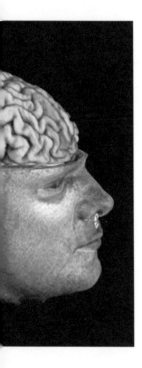

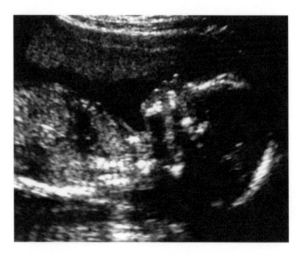

The Himalayas (above) and Karman Vortices (below).

5.07 Two-dimensional ultrasound image of baby in womb.

5.08 Seeing with sound: Sonar image of the *Empress of Ireland* wreckage in the St. Lawrence River, Canada.

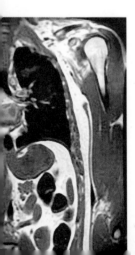

5.10 Infrared cameras detect electromagnetic frequencies below the range of visible light.

5.11 Nine billion images are produced yearly.

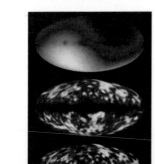

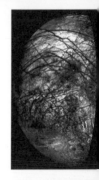

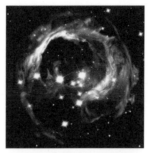

5.12 Multispectral telescope images from space.

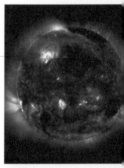

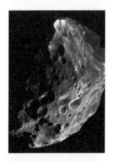
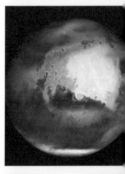

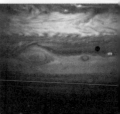

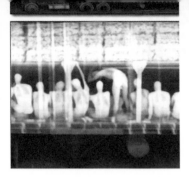

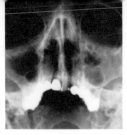

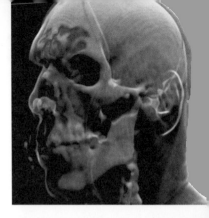

5.13 X-ray images: people smuggling and inside the human body.

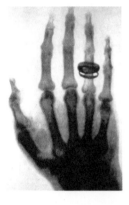

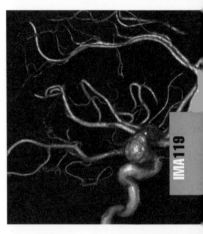

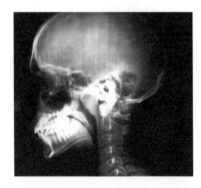

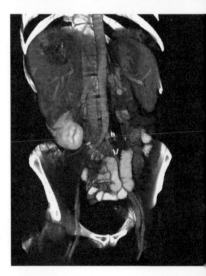

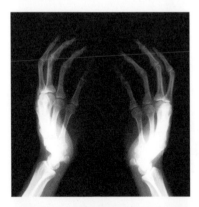

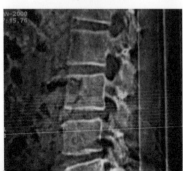

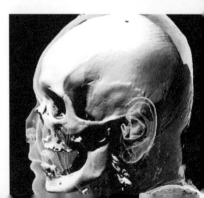

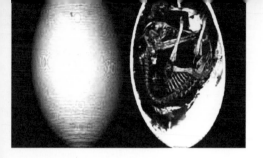

5.14 Computed Tomography (CT) images generate volumetric representations. This column: an emu egg, a regal horned lizard, a short-nosed fruit bat, and a pincushion protea.

5.15 Electron microscopes focus a beam of electrons with a magnifying power of 500,000 times the original size of such things as spider spinnerets, gold nanoparticles, and even a flea.

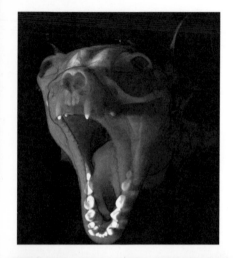

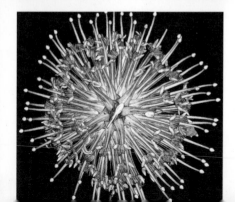

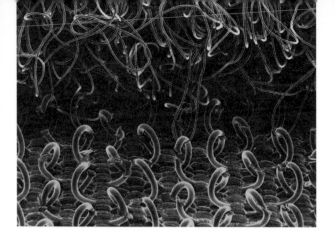

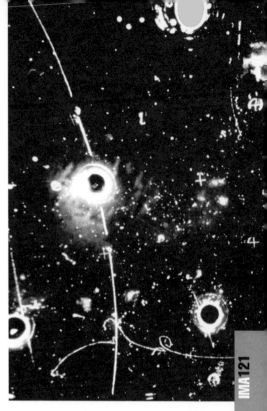

Electron micrography images by Dee Breger of Drexel University (this column): Velcro, a radiolarian, and iguana tail spines.

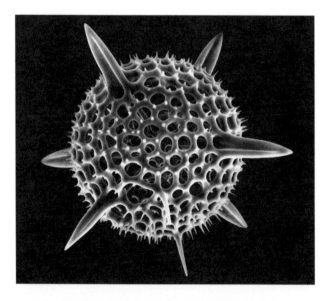

5.16 The first observation of neutral currents, from the Gargamelle heavy liquid bubble chamber at CERN, Geneva.

5.17 Field-ion microscope reveals individual atoms of tungsten that appear as bright spots organized in concentric rings.

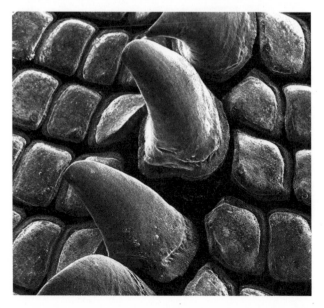

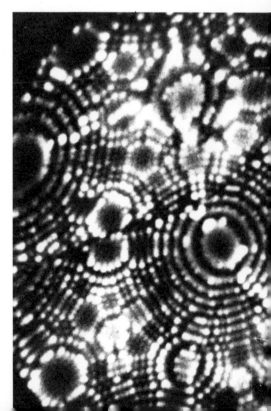

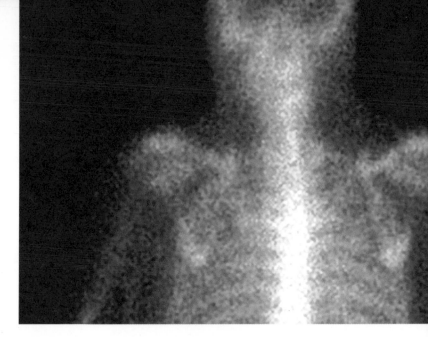

5.18 Bone in human torso is made visible via a scanning device that detects gamma radiation emitted by an ingested radioactive isotope.

5.19 Scanning tunneling microscope (STM) image of the quantum-mechanical interference patterns of 48 iron adatoms.

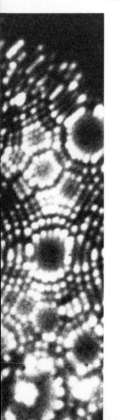

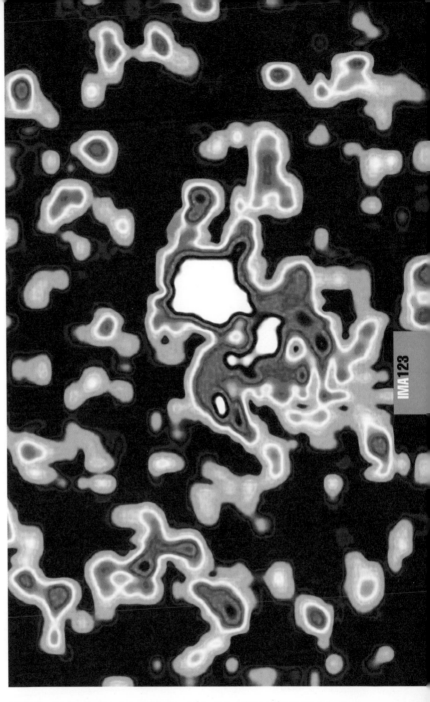

IMA123

5.20 A gamma ray burst, possibly caused by the collision of two neutron stars, releases in this singular event the energy that the Sun would generate over the course of its 10-billion year lifespan.

Felice Frankel
on micro- and
subatomic imaging

Mathematician Benoit Mandelbrot, the
father of fractal geometry, has called your
book, *Envisioning Science*, "priceless."
That's high praise.
I am so grateful for his support. Benoit has spoken
to me often about how the mathematics of his
fractals came after he first saw their images. This
is one of my soapboxes: Seeing a thing first often
leads to an expansion of ideas after.

We are visual creatures after all, are we not?
Absolutely. We have to be very careful because
there's so much visual noise out there. Digital
manipulation can change the information in
images, so we need to become intelligent about
how images are made. *Envisioning Science* is an
attempt to encourage future scientists, and also
nonscientists, to look at science through a camera's lens – to go deeply into the ideas that one
can capture in a photograph, or any sort of illustration. It's about visually communicating ideas
in science. That's what I'm totally committed to.

You did a wonderful job with the yeast colony,
for example.
Thank you. Yes, the hardcover version of
Envisioning Science has a detail of what some
people see as a stunning picture of a yeast colony,
which looks just like a flower. The patterns within
this flower are even more exciting. I took this
photograph in [yeast genetics pioneer] Gerald

Fink's lab at MIT for a science journal. I made
the picture with the petri dish and later digitally
removed it because I thought it would bring
more attention to what was going on in the most
important part of the image – these wonderful
patterns. We did, by the way, get the cover of
Science, which I was delighted about, but the
researchers were concerned that by digitally alter-
ing the image, I was also deleting information
about scale. It's those conversations that are so
important to have in the science community.

In the seventeenth century, natural philosopher
Robert Hooke suggested that nature should be
captured by "a sincere hand and a faithful eye
to examine and to record the things themselves
as they appear." Where do you draw the line
with the aesthetics of the scientific image?
The problem is, that line keeps moving around!
It really depends upon the purpose of the image.
If I make the picture for a submission to a science
journal, obviously we want to be as straightforward as possible. I don't think I'm bringing in an
aesthetic. I believe I'm revealing an aesthetic that
makes the image more accessible.

Tell me about a time when a scientist was
absolutely thrilled with the colorful portrayal
of his or her research.
Moungi Bawendi, a chemist at MIT who works
with nanocrystals, is making material that
fluoresces when you look at it under UV light –
you get different colors depending upon the size
of the crystals. He and his colleagues had photo-
graphed the nanocrystals in a solution with various
kinds of containers, so the different shapes got in
the way and distracted your eye away from the
colors of the nanocrystals, which was the important point of the science. Very simply, all I decided
to do was to make a picture of these beautiful colors lined up in similar vials with the view to communicate the science succinctly: As you change the
size of the nanocrystals, you change the color.
That's all we wanted to say. Moungi agreed.

You focus on science in the laboratory. Are you
equally amazed by the images satellites are
capturing these days?
How can you not be? These images are stunningly
beautiful. But you must remember that the colors
are produced digitally. The information is captured
through various filtration methods and by knowing
what filter has captured what cloud, for example,
the person working on the image will color it accordingly. This is not deceptive. NASA is doing a fabu-
lous job of bringing attention to the glorious world

out there. Chemists and physicists now need to do the same for the micro and subatomic world. It is just as beautiful and just as exciting as the cosmos.

How did you begin working with Harvard chemist George Whitesides?

Pure luck! I was on a mid-career (Loeb) Fellowship at Harvard and ended up auditing one of his molecular biology classes. I enjoyed his visual sensibility and invited myself to his lab after class one day. He introduced me to one of his postdocs, Nick Abbott, who was working on a submission to *Science* magazine. I looked at the photographs of the work and knew I could do better. (These were 4-mm-wide square drops of water that stayed within a grid pattern on the surface of a silicon chip.) To make a long story short, I asked them for all kinds of different material and went to work. Not only did I get the picture in focus, but we also made the cover of *Science*, which was quite remarkable. That was the beginning.

5.21 Felice Frankel's colored scanning electron microscope image of nanowires.

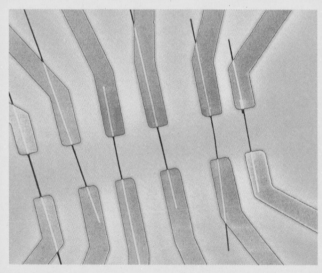

When you work at the atomic level, do you make images using electrons and atomic forces?

I've worked at the nano level, a bit larger than the atomic level. And I always work with people who know much more about the equipment than I do. These instruments require extensive training, so I sit with either the technician or the researcher and ask them to do this or that to the image, to represent it the way I would like to. This whole realm is very difficult to communicate, even for students in the field. One of the researchers was telling me that he was with an undergraduate who was holding a test tube of nanotubes up to the light and proclaimed, "I don't see them." Well, of course you can't see them with the naked eye, because you need electrons to see them!

Is two-dimensional photography limited in an era of moving images and multidimensional computer modeling?

I firmly believe that there is a strong place for the still image. I work with animators and people working with dynamic imaging, but I wonder if we get more information from a moving animation. Take, for example, the series of still pictures I made over time of a Belousov-Zhabotinsky reaction. The reaction is continuous, but if you look at stills taken every eleven seconds and display them within a grid of twelve images, you can actually see how one moment changes to the next. If I were to animate the process, would you get more information by seeing it in motion? We can be blown away by knock-your-socks-off animation, but sometimes it happens too quickly to sufficiently communicate deep ideas.

What images in science would you say have over time changed the way we see ourselves?

The image from space of the earth rising above the lunar surface. The double helix. The X-ray. And then of course there are pervasive images in science that are dead wrong, like the dumb illustration of evolution, where an ape-like figure slowly stands upright. This is not what evolution is about. It's a terrible simplification of something much more complicated.

In the introduction to *Envisioning Science*, the late author and educator Phylis Morrison traces the history of the scientific image back to Neolithic cave paintings in France and early sketches by Copernicus and Darwin. How broadly do you define the scientific image?

Yes, what is a science image? It's a very good question. People painted images in the Chauvet cave 30,000 years ago. Although they didn't know they were making a science image, they documented bison and whatever else they were seeing. I document what I see too. I feel that the world is science, and if one takes a picture of part of the world, then that picture is a representation of that world, and so it is a science picture.

Felice Frankel is a research scientist and science photographer at the Massachusetts Institute of Technology in Cambridge, Massachusetts.

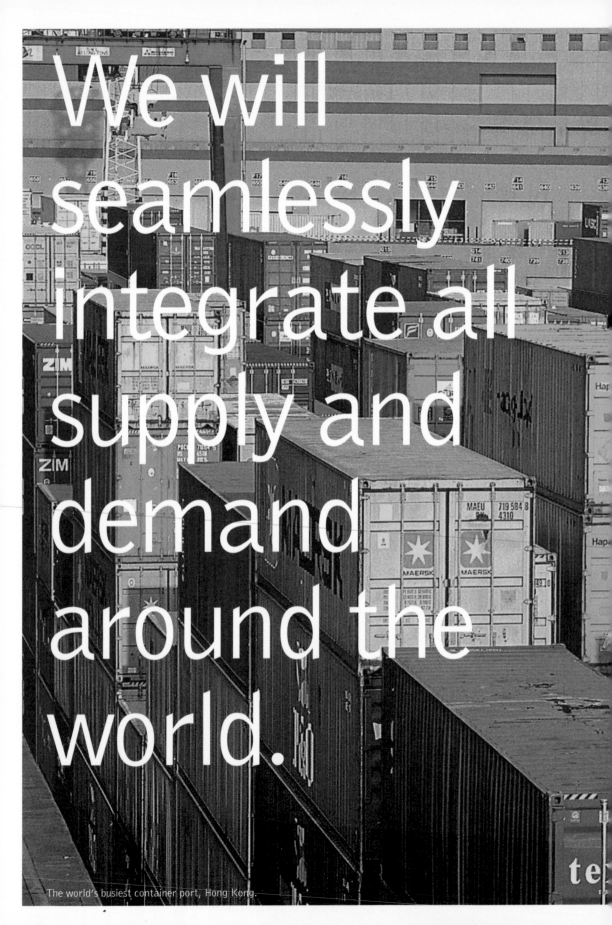

We will seamlessly integrate all supply and demand around the world.

The world's busiest container port, Hong Kong.

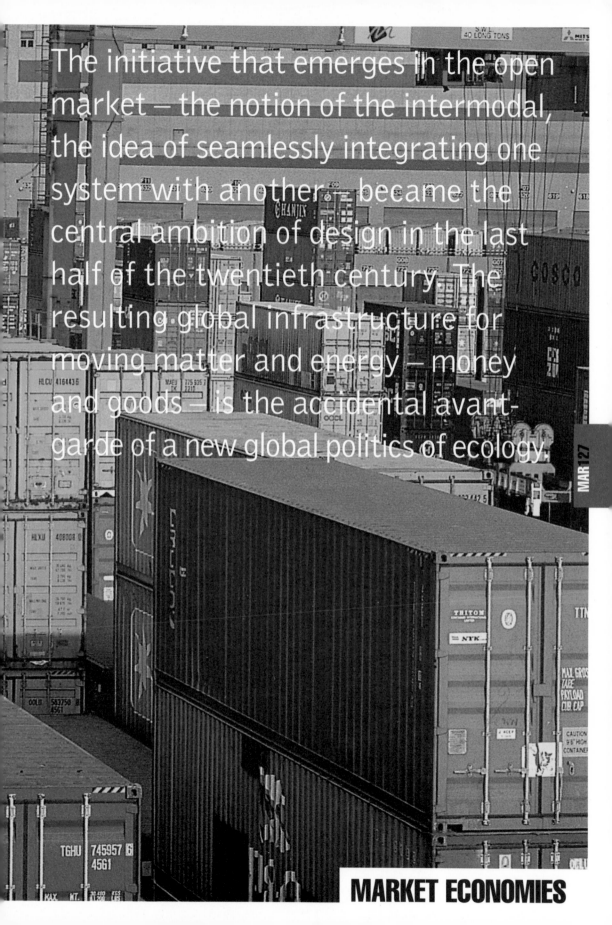

The initiative that emerges in the open market – the notion of the intermodal, the idea of seamlessly integrating one system with another – became the central ambition of design in the last half of the twentieth century. The resulting global infrastructure for moving matter and energy – money and goods – is the accidental avant-garde of a new global politics of ecology.

MARKET ECONOMIES

6.01

6.02

BEHIND THE SCENES. High-speed conveyor belts and state-of-the-art technology help move merchandise efficiently through the Wal-Mart distribution centers (6.01), keeping nearly 3,000 stores in stock. Handling more than 3.3 million packages daily requires the largest all-cargo fleet in the industry. The FedEx Express fleet, at the main hub in Memphis, Tennessee (6.02), includes more than 650 aircraft.

0123456 7890123

TINY TRACKERS. Since bar codes (6.03) first appeared in the 1970s on a package of Wrigley's chewing gum, they have revolutionized shopping and retail logistics and become the global language of business. The next threshold for retail logistics is Radio Frequency Identification (RFID) tags (6.04). These are small electronic assemblies in which information can be stored, read, even rewritten. Although these can support "intelligent" functions for the store of the future, they raise obvious concerns about personal privacy.

Integrated systems: When everything is connected to everything else, for better or worse, everything matters.

When a Supercenter moves into town, competitors often are wiped out.

– "The Wal-Mart Effect," *Los Angeles Times*, November 23, 2003

Wal-Mart is the retail behemoth that it is today because of the integrated supply-chain system that supports its infrastructure. Made up of electronic data interchange networks and an extranet used by Wal-Mart buyers and 10,000 suppliers, it culls information about sales and inventory levels in every store. The extranet's database holds more than 100 terabytes of data – the equivalent of more than five times the entire contents of the U.S. Library of Congress. Wal-Mart has stores in more than ten countries including nearly 3,200 outlets in the United States alone. Based on revenues – $244.6 billion in sales in 2002 – it is the world's biggest company. It is also the largest employer in the world, with 1.3 million employees, and plans to hire 800,000 more over the next five years. Viewed as a piece of the global economy, Wal-Mart is the nineteenth-largest economy in the world. Its sales on the day after Thanksgiving in November 2002 were $1.42 billion. This means that Wal-Mart's revenue on one single day was larger than the annual gross domestic product (GDP) of 36 separate countries. It has more people in uniform than the U.S. Army, and if the estimated $2 billion it loses through theft each year were incorporated as a business, Wal-Mart would rank No. 694 on the Fortune 1,000.

The effect of this scale of integration, besides the obvious competitive impact, is to deliver more for less. The relentless search for value in turn drives global integration, connecting market to market, culture to culture. As a consequence, the average Wal-Mart consumer can purchase goods that would have sold for $100 ten years ago for only $25 today.

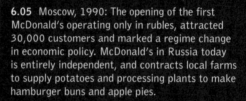

6.05 Moscow, 1990: The opening of the first McDonald's operating only in rubles, attracted 30,000 customers and marked a regime change in economic policy. McDonald's in Russia today is entirely independent, and contracts local farms to supply potatoes and processing plants to make hamburger buns and apple pies.

6.06 Residing outside the gates to the Palace of Heavenly Purity in the complex of the world's largest imperial palace in Beijing now sits the world's tiniest Starbucks coffee shop. Although the ancient home to China's emperors was declared a UN World Heritage site, not even The Forbidden City is immune to commercialism.

Corporate accountability: The Home Depots and Nikes of the world have greater capacity to achieve more for greater good because of their scale. One incremental change for them becomes massive change for the entire industry.

The Natural Step pedagogy has proven itself to be among the most effective ways to establish a foundation for the mind-set shift needed for the twenty-first century enterprises to work.

– Peter Senge, author, The Fifth Discipline

Love or hate Starbucks, when this corporate giant ensures 1% of its coffee is fair trade, a lot of money ends up going toward improving the conditions for agricultural workers in the coffee industry.

The Natural Step is a nonprofit organization that helps some of the largest resource users, like Starbucks, examine their entire business system through the lens of sustainability. Since forming a relationship with The Natural Step in 1998, Nike has made shoeboxes 10% lighter (saving 4,000 tons of raw materials and $1.6 million annually) and switched to water-based cements in 90% of its shoes (saving more than 1.6 million gallons of solvents per year, the equivalent of more than 32,000 barrels of oil). McDonald's Sweden now runs 75% of its 233 stores on renewable energy, serves organic dairy products, and recycles 90% of all restaurant waste. The Home Depot has implemented a new wood purchasing policy that states it will give preference to wood originating from certified, well-managed forests wherever feasible and will eliminate purchases from endangered regions. Because it is the world's largest home improvement retailer, Home Depot has the power to insist that the industry accommodate this policy. No neighborhood hardware store could ever do this on its own.

When the big guys say, "We're going to change things," they actually have the capacity to do it. Although bigger is not always better – and it certainly runs counter to the current anticorporate tone – the most poignant possibilities for change exist here.

Catherine Gray
on sustainable
business

What is your main focus at The Natural Step?
Using a scientific framework, it is our mission
to move society towards sustainability. To do
this, we focus primarily on large multinational
companies because they have the leverage and
the ability to significantly impact ecosystems, as
well as their supply chain. Large retail brands
have one foot firmly in ecosystem services and
one foot firmly in a brand that is very well known
with consumers.

**How do you typically work with the Fortune
500 companies?**
We work in partnership with senior teams of
companies to help them understand the science
beneath sustainability and how to build an
integrated approach deep into their strategy
and operation. We use a combination of basic
education and awareness building combined with
business analysis, from a sustainability perspective, so that they understand where their
biggest impacts are, where their biggest risks
are, and where their biggest opportunities
are. Once they have that picture, we help them
build a vision and a strategy to begin to move
in a more sustainable direction. All of this is
informed by their business context, what's
important to them, what's happening in their
company at the point that we enter, and how
we can begin to help them get a bigger return
on their bottom line.

**You must have on your team of researchers and
experts then, both scientists and strategists
from business?**
That's correct. The Natural Step is both a scientific think tank and business consulting group.
We have Ph.D. biophysicists and social scientists
combined with people who've come right out of
leading management consulting groups.

**How would the world change if more companies
were to incorporate the principles of The
Natural Step into their practices?**
At a fundamental level, our industrial system was
built without sustainability in mind. It just wasn't
an imperative like it is today. Nature was considered limitless and waste was a non-issue. But
we're now at a point in our evolution where we
understand these issues to be critical. So if every
company began to build this awareness into their
systems, we'd have a very different world. Right
now, every single life-supporting system on our
planet is in decline: fresh air, clean water, rich
topsoil, and productive forests. The real game on
the planet right now is working toward reversing
this trend and moving toward sustainability.

What are the barriers to corporate sustainability?
There are a number. We're still in the very early
days of this, and if you look at our current economic and political context, sustainability is not
incentivized. In fact, the reward system currently
in place supports unsustainable business practices.
So it's an uphill battle, especially if you're a big
multinational company and the kind of change
you're talking about is deep and long term. In a
culture of quarterly profits, we're looking at how
we can get the short-term wins as well as hold
that long-term vision in place. It's a tricky balance.

**How extensive is the Natural Step's global
network?**
We have offices in twelve countries now, with
about sixty people working full-time, and hundreds of academic partners and scientists. In
each country that we have an office, while we
have a core methodology, it's rolling out very dif-
ferently because the issues are different in each
context. In our South African office, for example,
there's a heavy focus on the social side of sustainability. In Japan, there's a heavy focus on persistent chemicals and compounds. In the United
States, the focus has been on large companies. In
New Zealand, the focus is on smaller companies.

**What significant changes has The Home Depot
made since working with you?**

A few years ago, The Home Depot began to take a hard look at a number of issues around sustainability, especially old growth timber. Arthur Blank, the CEO at the time, was very concerned and astute around environmental issues and did not want to leave a legacy of environmental destruction behind. The Home Depot was also under attack by a number of activist groups for its wood purchasing policies. So it began to look at where it was sourcing its wood and sought out certified woods. We did a two-year thorough analysis with The Home Depot and began educating its senior team, merchants, and key staff in this framework of sustainability. We developed a roadmap for the company to move forward, taking into account everything from its paints to its plastics to its garden supplies.

Nike is always a hot target for public scrutiny. How have you helped?

Nike entered the conversation of sustainability over ten years ago. Like The Home Depot, it had a campaign aimed at it with respect to its labor practices. But it already had in place a team focused on environmental and social issues. Nike has huge impacts, both environmental and social, down the supply chain, and recognizes this. We helped 75 of Nike's top designers and senior representatives innovate around sustainability. As a result, Nike has come up with a whole host of new designs. The company has taken a hard look at the persistent compounds and heavy metals in its products.

What about Ikea?

We worked with Ikea in Sweden in the early days of our organization. The CEO woke up one morning with the unfortunate situation of reporters at his front door asking what he was going to do about the off-gassing of the formaldehyde in their bookcases. The CEO had no idea what to do and asked The Natural Step to come in and help educate him and his staff in sustainability and give them a framework to help them make better decisions. We analyzed their systems and looked at everything from the formaldehyde in their bookcases to their paints, to their wood, and on and on. As a result, Ikea is a leading company in the field of sustainability and has been a key driver in the realm of endangered forests working with significant noprofits, like the World Wildlife Fund.

And Starbucks?

We were asked to come in and help them understand their environmental footprint. Working side by side with their team, we analyzed their system from a sustainability perspective, looking at every-thing from the pesticides that are used to grow coffee to some of the fair trade practices and social conditions that surround the coffee industry, to transportation and energy. From there, we developed the metrics for them to begin to measure their impact and improvement over the course of the year.

You also work with McDonald's, don't you?

We started very early on with McDonald's in Sweden. Ten years ago, McDonald's was one of the most loathed brands in Sweden. The CEO at the time thought this was unacceptable and wanted to do something about it. So he brought us in and we trained the employees in sustainability in The Natural Step framework. If you go into a McDonald's in Sweden today, you'll find 100% organic dairy products, as much organic beef as the market can supply, as many organic vegetables as the market can supply, and no genetically modified organisms in the food. Many of its stores are powered by renewable energy, solar as well as wind and some natural gas. It took a hard look at its transportation and moved to transporting its goods by train and biofuel cars. It has incentive programs for its employees to car pool. They've achieved a great deal in terms of cost savings and waste savings and employee morale. Over three years, McDonald's went from one of the most loathed brands in Sweden to the third most popular brand in the country. Today we are working with McDonald's globally to help it build an integrated approach to sustainability.

Tell me about *Ants, Galileo, and Gandhi*.

This is a book that was inspired by one of our conferences on sustainability. It's a wonderful resource that highlights what's happening across various sectors today, from start-up companies to multinationals. The title says it all. When you think about sustainability, it's going to take the coordination of ants, the scientific vision of Galileo, and the conviction and compassion of Gandhi for us to head in this direction as a society.

Catherine Gray is the president of The Natural Step, an international research and advisory group.

6.07 A FLOCK OF VISAS. A "chaord" is a term invented by Dee Hock that refers to the zone between chaos and order, where living systems thrive. It is any organism, organization, or system that is self-organizing, self-governing, adaptive, nonlinear, and complex.

One economy, one ecology: Imagine the possibilities of an entity that circulates globally and guarantees monetary information in the form of arranged electronic particles. Herein lies the brilliance of Dee Hock and the resilience of chaordic systems.

Visa has elements of Jeffersonian democracy, it has elements of the free market, of government franchising – almost every kind of organization you can think about. But it's none of them. Like the body, the brain, and the biosphere, it's largely self-organizing.
– Dee Hock, founder and chief executive officer emeritus of Visa International and Visa USA

Visa International, the corporation whose product is coordination, came to be in 1970 when a Seattle-based banker named Dee Hock codified and applied the dynamic principles of chaos theory to the BankAmericard system. His ideas of decentralized collaboration within a single organization came right at the time when digital and electronic transactions were on the ascension. It took several years to evolve this "chaordic" (chaotic and ordered) structure, and many doubted throughout its development that it was ever going to work; but Hock persisted. In the early stages of design, he looked to the resilience of weather and ecological systems, placed a harmonious blend of competition and cooperation at the organization's root, and hoped that it too would stand the test of conditional changes. Visa now consists of 20,000 financial institutions, 14 million merchants, and 220 participating countries and territories that serve more than 600 million people. Hock proved hands down that organizations work best when built on biological concepts and metaphors.

In 1984, Dee Hock retired from Visa. Thirteen years later, with the support of major foundations, he founded The Chaordic Commons of Terra Civitas, a not-for-profit institution. His ambition is to link individuals and organizations throughout the world in a concerted effort to develop, disseminate, and implement more effective and equitable concepts of commercial, political, and social organization.

Hazel Henderson
on global markets

How is the field of economics dealing with technological change?
Part of the thesis in most of my books critiquing the traditional economics is that they missed the most important driving variable in the whole economic process – the evolution of technology and the unfolding of the Industrial Revolution itself. Which is really all about change. Economic theory considers technology as a given. This is why economics, I've always said, is backing into the future looking through the rearview mirror.

Both Marshall McLuhan and Bruce Sterling have said that a good futurist is one who can predict the present.
I think that's a good way of saying it. There's another thing about being a futurist, and it relates to personal responsibility for the future. In other words, we are all making the future every minute that we live, by way of our collective and individual decisions. If we think of it like that, everybody is really a futurist.

Tell me about your Layer Cake with Icing.
This is one of my earliest diagrams. I use a layer cake as a metaphor for a total productive system of an industrial society. If you can visualize it, the icing on the top is the private sector, which rests on the layer below, the public sector. These top two layers are the only ones economists typically measure. But in my analysis, there are two lower layers that are non-monetized and invisible to

economists, but which are really supporting the whole thing. These include the Love Economy – unpaid productive work like raising children and maintaining the household, serving on the school board, do-it-yourself housing, rehab – and Mother Nature, the vast wealth of biodiversity that keeps our air and water clean and provides all the food and fibre and resources we need to sustain life, which go completely uncounted. When an economic system doesn't take into consideration these two vital lower layers, which support the official money economy, then both the society and ecosystem get kind of cannibalized. Wall Street and the financial community all over the world are really living in a fool's paradise.

How much will it cost for every country in the world to shift to sustainable development?
It was estimated at the summit in 1992 in Rio de Janeiro that it would cost somewhere around $650 billion. But the thing is, right now those same governments are spending well over $1 trillion a year on unsustainability, which subsidizes fossil fuels, nuclear power, and high-technology approaches to agriculture. If governments would simply stop subsidizing unsustainability, we could move to sustainability and have about $350 billion left over each year.

Would it require a massive overhaul of global infrastructure?
No, it isn't really an overhaul of infrastructure so much as a removal of the suppressive political influence on the economy. Take the regime change in Brazil, for example. President [Luiz Inácio] Lula [da Silva] is interested in sustainability, and so you see Petrobras, the big oil company, suddenly developing hydrogen, wind, and solar divisions. The same thing is also happening with British Petroleum and Shell. Even OPEC is beginning to realize that its future is not so much in oil as it is in investing in these new technologies. The pharmaceutical industry is another interesting sector undergoing change. It is beginning to realize that it cannot extort these enormous prices for its antiretroviral drugs in countries like South Africa or Brazil. In Brazil, they simply make their own; they have their own industry, and that's why they've done a much better job dealing with HIV/AIDS than countries that have to rely on costly pharmaceuticals from the big companies.

More than the big companies, it seems that nongovernmental organizations are increasingly harbingers of social change. Why is that?

Because they don't have these vested interests. The NGO movements around the world, as I have known them over the past thirty years, start up as a result of being on the receiving end of all unsavory social impacts. As long as companies are permitted to externalize the social and environmental costs from their balance sheet, NGOs are going to organize and try to push

communities that are completely sidelined from traditional banking and economic networks to match their own needs and resources, to create sustainable livelihoods outside of the money circuits. I am involved in a start-up in London right now that provides barter and exchange services to NGOs, at www.via3.net.

I think we can globalize human rights; we can globalize more transparent and democratic societies. We can globalize the equality of women. We can globalize environmental protection.

those costs, quite rightly, back onto the corporate balance sheet.

It reminds me of Alvin Toffler's *The Third Wave*, in which he introduces this notion of a producing consumer. How are corporate goals changing as a result of the "prosumer"?
Many corporations have admitted that if the NGO groups didn't challenge and push them, they couldn't make certain changes because Wall Street and the financial markets expect them, according to the old model, to maximize shareholder returns. We're all trying to move towards the stakeholder model, away from the stockholder model. In other words, a company really has responsibility not only to its investors and shareholders but also to its consumers, its employees, the community, the environment, and future generations. This conversation is alive and well and a lot of corporate groups feel the pressure to change. And frankly, I'd like to increase that pressure.

What is the significance of barter here?
Because I focus on the nonmoney side of production and transactions in my books – all the cooperative, unpaid activities that are the bedrock of all of the world's economies – I've naturally been interested in explaining barter. In the world today, where we have six billion people sharing the planet, there are still about two billion people that will never see the inside of a bank or even get a microloan from Women's World Banking or ACCION or the Grameen Bank or any other of these microlenders. And so, barter is so important because it allows

You once described a scenario where two big freightliners passed in the night, one from Tokyo heading to the U.S. and the other leaving the U.S. heading to Japan. Absurdly, both were carrying cars.
Absolutely! I think most countries in the world would be better off creating homegrown economies. Anything that they could produce in their own society, they would develop domestic markets for. And only those things that really were unique in terms of minerals or whatever would be part of world trade. What makes sense to trade internationally, of course, are ideas. The famous economist John Maynard Keynes once said it was much better to ship around recipes than it was to ship around cakes and biscuits. This, to me, is the future, where we have a globalization system based on being able to savor all of the world's best ideas – the world's best music, food, art, recipes, and clean, green technologies. This is the direction I think we should be heading in world trade.

What are the positive potentials of globalization?
I think we can globalize human rights; we can globalize more transparent and democratic societies. We can globalize the equality of women and environmental protection. We can reorganize all of the new forces – this is a transition going on now, as we've noticed, with NGOs, the United Nations and many of its humanitarian and development agencies, and the ILO (International Labor Organization) and the WHO (World Health Organization). Together these make up a new global force in the world. Without question, the new world superpower is world public opinion. When that's linked to mass media, I think we can end run a lot of these old structures and transition more effectively.

Hazel Henderson is an evolutionary economist based in Florida.

Prosumer: With the proliferation of digital networks the world over, the electronic marketplace has gone from empowering the consumer to supporting a global civic society. Power to the people.

The new electronic independence re-creates the world in the image of a global village.

– Marshall McLuhan

In *The Third Wave*, Alvin Toffler predicted the emergence of a producing consumer – a "prosumer" – out of the transition from the Industrial to the Information Revolution. Advances in technology have indeed given capacity to the consumer. McLuhan said the photocopier makes everyone a publisher. Now, with digital video editing software, everyone's a filmmaker; with Flash, everyone's an animator; and with eBay, everyone's both a buyer and a seller. The Information Age has also reintroduced nonmoney exchange to where markets were already flourishing. Indigenous communities have proven for thousands of years that barter is efficient. Governments commonly barter: India makes exchange agreements with China and Russia for power plants, heavy machinery, oil, and trucks. Corporations barter, too – over trillions of dollars' worth of services: bandwidth, airline seats, hotel rooms. Along these same electronic circuits, citizens of all nations are making nonmoney exchange agreements towards a unified social economy; a worldwide civic society is rising up through the cracks of corporate globalization.

Booker Prize–winning author Arundhati Roy addressed the opening Plenary session of the World Social Forum in Mumbai in January 2004 with this: "…as long as our 'markets' are open, as long as corporations like Enron, Bechtel, Halliburton, Arthur Andersen are given a free hand, our 'democratically elected' leaders can fearlessly blur the lines between democracy, majoritarianism, and fascism. Radical change will not be negotiated by governments; it can only be enforced by people."

6.08 AS EASY AS EBAY. EBay enables the at-home consumer to engage in online barter with a trading partner anywhere in the world. Everything you could ever wish for, going once, going twice, sold to the highest bidder.

We will build intelligence into materials and liberate form from matter.

Bio-polymer scaffold seeded with liver cells.

Material has traditionally been something to which design is applied. New methods in the fields of nanotechnology have rendered material as the object of design development. Instead of designing a thing, we design a designing thing. In the process, we have created superhero materials and collapsed the age-old boundary between the image and the object, rendering mutable the object itself.

MATERIAL ECONOMIES

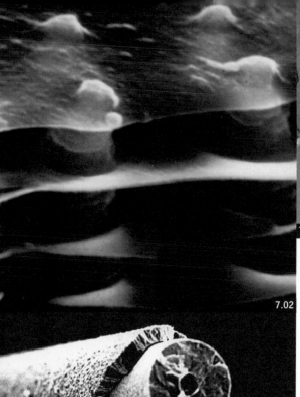

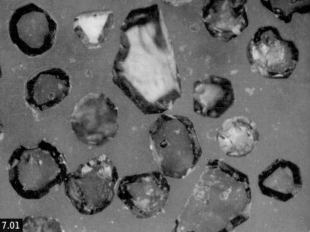

7.01

7.01 cBN. Cubic boron nitride crystals, measuring 500 microns, are synthesized at the Institute for High Pressure Physics, Russian Academy of Sciences. The high pressures and temperatures of fluid catalyst systems turn these into powders, which are used for making hardened steels and ceramic- and metal-bonded grinding wheels.

7.02

7.02 NACRE. Materials scientist Mehmet Sarikaya, of the University of Washington, takes his inspiration from hard tissues in nature for the design of ceramics at the nanometer scale. Specifically, he mimics nacre, (mother of pearl) from mollusk shells, since its micro-architecture is the result of an evolutionary design for an ideal impact-resistant material providing armor to the mollusk.

7.03

7.03 and **7.04** CVD. At the University of Bristol School of Chemistry, the researchers study chemical vapor deposition of diamond-thin films on substrate materials. Future applications for CVD diamond include cutting tools, thermal management, optics, electronic devices, electron emitters in flat-panel displays, electrochemical sensors, particle detectors, and micromechanical devices and sensors.

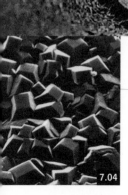

7.04

7.05 B_6O. This grain of boron suboxide, seen here in a pseudocolor SEM (scanning electron microscope) image, was synthesized at the Arizona State University (ASU) Materials Research Science and Engineering Center (MRSEC). The scientists took boron and boron oxide and heated the mixture at high pressure to form orange-red particles. Boron suboxide ranks as the third-hardest substance in the world; only diamonds and cubic boron nitride are harder.

Superhard: Materials scientists use high-pressure physics and nanotechnology to design synthetic materials that mimic Nature's beauties, such as diamond and mother-of-pearl, which are harder and tougher than the rest.

The central challenge to modern materials science is the rational design and synthesis of new materials with exceptional properties.

– David M. Teter, geochemist

As we continue to develop tools and knowledge in the realms of quantum physics, nanotechnology, imaging technology, chemistry, biology, and molecular engineering, we are better able to design materials to meet human – and increasingly superhuman – needs. Scientists are also looking more and more to nature for blueprints and patterns on which to base the design of synthetic materials. The knowledge gained from looking to the atomic level of natural structures is used to enhance existing synthetic materials and develop new supermaterials.

When it comes to hardness, there is no theoretical definition in materials science. Various scales measure superhard materials, with diamond as the standard by which all others are assessed. The oldest is Mohs' hardness scale, which was first devised in 1812 by German mineralogist Friedrich Mohs (1773-1839). Mohs developed a system of ten readily available minerals – talc, gypsum, calcite, fluorite, apatite, orthoclase, quartz, topaz, corundum, and diamond – to determine a material's relative hardness by how well it resisted scratching by another material. To this day, no material can outscratch diamond, and field geologists use Mohs' scale and the following mnemonic device to remember the intended order of things: Those Girls Can Flirt And Other Queer Things Can Do.

Modifications to the list have since been made to incorporate additional substances, such as garnet, fused zirconia, and boron carbide. And diamond, still the hardest, could be stripped yet of its rank. The current contenders, in decreasing order of hardness, are cubic boron nitride (cBN) and boron suboxide (B_6O).

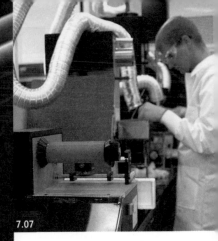

7.07 and **7.08** BIOSTEEL.
In an effort to mass-produce synthetic-spider-silk-performance fibers, the U.S. Army Soldier Biological Chemical Command and Nexia Biotechnologies, Inc. in Quebec have inserted an orb-weaving-spider gene into the mammary glands of dwarf goats from West Africa; this strain of goats naturally Breeds Early and Lactates Early (BELE). The result is BioSteel, silk you can milk.

7.06

7.06 EDIBLE SILK. A Scanning Electron Microscope (SEM) image of dragline silk of *Nephila edulis*, the edible golden silk spider. It tastes like peanut butter, but "without the objectionable consistency," says Mike Robinson, the former director of the Smithsonian Institution National Zoo.

7.08

7.09 GECKO FEET. Geckos scurry up walls using van der Waals forces, the intermolecular attraction between the atoms of the setae – the millions of tiny keratin hairs that cover gecko feet – and the atoms of the climbing surface. These forces help geckos tackle even the most slippery of surfaces and suspend from glass with just one toe.

7.11

GECKO TAPE. Scanning Electron Microscope (SEM) image showing the spatulae on a single gecko seta (7.10). Millions of synthetic setae cover a square centimeter of microfabricated gecko foot hair (7.11).

7.10

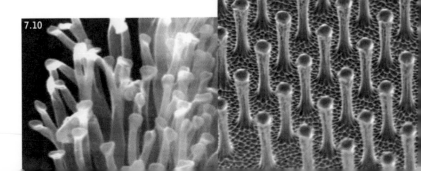

Superstrong: The U.S. Army, bio-technology companies, zoologists, biologists, and many more are racing to develop synthetic fiber to surpass the support strength of spider silk and synthetic adhesive to outperform the grip strength of gecko feet.

Mimicking spider silk properties has been the holy grail of materials science for a very long time.

– Jeffrey Turner, president and CEO of Nexia Biotechnologies, Inc.

Many scientists today are closely examining the spinnerets of orb-weaving spiders and the micro-hairs of gecko feet for inspiration in the design development of superstrong fibers and adhesives, respectively. The powerful synergy of synthetic materials modeled on spider dragline silk and the microscopic foot-hairs that give geckos their grip could endow us all with sufficient strength to scale walls and suspend our own weight from the ceiling using only our palms.

For more than two decades, Fritz Vollrath of Oxford University's Department of Zoology has studied *Nephila clavipes*, the golden silk spider that produces silk with a tensile strength that far outperforms steel, per unit weight. Vollrath has calculated that webs spun to scale by a human-size spider would be as big as a football field, and orb webs with finger thickness could catch a jumbo jet at landing speed, yet be folded up entirely to fit into a tea chest. Vollrath and colleague David Knight have developed a new company, Spinox, whose aim is to devise novel ways to mimic spiders' ability to spin silks for sutures, surgical implants, and protective clothing.

Andre Geim and colleagues at the U.K.'s University of Manchester have designed microfabricated gecko foot hair using a material called Kapton, which has identical measurements to gecko hairs: 2.0 microns high and 0.2 microns in diameter. A 1-cm square piece of this "gecko tape" has about 100 million synthetic setae and the strength to support a weight of 1 kg.

Philip Ball
on materials science

Explain this idea of "the material is the mechanism."
It's about moving away from the classical idea of materials – inert stuff that serves a structural role – toward the more contemporary notion of materials. More and more, materials are active and respond to stimuli in their environment. Materials can light up when an electric current is passed through them; materials can swell and contract in response to changes in temperature or acidity. Increasingly, there's a blurring of boundaries between what is a material and what is a machine.

If you were to draw a materials family tree, where would you begin and what would be its main branches?
The main branches are ceramics (including rocks), which is the oldest branch of materials; natural materials (wood, leather, plant fibers, etc.), also very old; metals; and synthetic polymers. Since the beginning of the twentieth century, it would be fair to say that the introduction of synthetic polymers has been the biggest change we've seen in materials science. Things now, of course, are very diverse. The branch tips have split into countless categories, many of them overlapping. But one of the most significant has to be semiconductors.

Which natural materials have changed our lives or have the hope to change our future?

In addition to the ones I've already mentioned, there's paper, which enabled the printing revolution. Looking ahead, we're starting to explore nature more closely and use its principles to make new types of materials. Biological materials are made from either protein, where the raw materials are amino acids, or nucleic acids like DNA and RNA, or polysaccharides (carbohydrates), where the raw materials are sugars. Biology manages to do an incredibly wide range of things with proteins, in particular [create] horn, skin, tendon, and transparent material that makes up the lenses of our eyes. So materials scientists are inspired to look to proteins to see how they might be able to redesign them. Genetically engineered bacteria, for example, can produce new kinds of proteins that might create novel and biodegradable plastics.

What about the metals family?
There is still plenty of work going on in good-old-fashioned metallurgy, although it's certainly not old-fashioned anymore. Nowadays, some researchers use computer calculations to come up with promising combinations of metallic elements. There's also the combinatorial method, where an entire "library" of new materials is produced: an array of combinations of elements in very small quantities. We then have to develop a way of testing the relevant properties of these substances to find those combinations that seem most promising. This automated approach to materials discovery has come about in the last ten years or so.

And polymers?
The earliest human-made polymers were derived from natural ones. Celluloid was made by chemically treating plant-derived cellulose. Vulcanized rubber was made by vulcanizing, or stiffening, the gum from rubber trees. These were the first semisynthetic polymers. In the early twentieth century, techniques for making more versatile polymers were developed, generally starting from the raw building blocks of the polymers' chainlike molecules. Rayon, which is also a semisynthetic cellulose-based polymer, was used for making fibers and textiles. Nylon came next, a hugely successful completely synthetic polymer. Polystyrene and polyethylene also were early examples of totally synthetic, petrochemical-based polymers.

Imagine putting together a superhero team of materials. What would the superlightweights be?
I'd go with polymers quite generally, and also with aerogels, which can be made from a whole range of materials. Aerogels consist of tiny

particles – these could be made of carbon, metals, or, most commonly, ceramic materials – joined together in a weblike three-dimensional structure and surrounded by air. They're mostly empty space. In fact, you can make aerogels that are 99% empty space! They're translucent and are often talked about as "frozen smoke." But they're also stiff, like a block of ice, and strong enough to support things standing on them.

How about the smart materials?

Shape-memory metal alloys and shape-memory polymers, since they remember their shape. You can take a straight wire made from Nitinol, a metal alloy made of nickel titanium, bend it into a shape, stick it in a cup of hot coffee, and the wire straightens out again. Shape-memory polymers change in response to light and have biomedical uses, as artificial muscles or self-tightening suture threads. Piezoelectrics are also smart materials. When squeezed, they produce an electric field. When voltage is put across them, they contract or expand. Piezoelectrics are used in loudspeaker drivers and microphones, which interconvert electric signals and acoustic signals (sound waves).

act as a mask for etching circuit patterns into the chip; the silicon gets etched away only in the places where you want it.

Which materials have superstrength?

Carbon nanotubes have the potential to be stronger and stiffer than just about any other material we know. These are like very, very narrow drinking straws made of carbon. Their width is typically about 10 nanometers or so, which is 10 millionths of a millimeter. That's tens of thousands of times narrower than a human hair. The current challenge is to grow them long enough to have practical application.

Is biomimetics at play in all of the above?

Biomimetics – learning tricks from nature – is currently a strong theme in materials science. In making tough materials, materials that don't break easily, the inspiration from nature is the mollusk shell – mother-of-pearl, or nacre, for example. This is a composite material, a mixture of sheets of mineral-like material and thin films of proteins. The combination of a hard and brittle material (ceramic) with a soft, organic material produces a very tough substance.

Since the beginning of the twentieth century, it would be fair to say that the introduction of synthetic polymers has been the biggest change we've seen in materials science.

And in the realm of the very, very small?

This is where the line between what's material and what's a device begins to blur. A nice example of what you can do at very small scales using simple, cheap chemistry is provided by self-assembled monolayers. These are very thin films of organic material that can be patterned at high spatial resolution and easily applied to the surface of gold or silica, on a silicon chip. One can make patterns that are smaller than what's used for shaping conventional microelectronic circuits, which is one of the big concerns of the information technology industry. So, rather than developing expensive methods for carving silicon chips into ever-smaller wires and transistors, you might use self-assembled monolayers. You can basically just stamp the surface of the chip with a rubber stamp covered with a kind of "ink" that forms a self-assembled monolayer on the chip. These films

When IBM famously manipulated atoms to spell out its corporate logo, what was the significance?

It showed us very vividly how much control we have over matter at the scale of individual atoms. There's a big question about whether the way they did that is really going to be useful in building materials, however. I think that one can probably get a lot further by using clever chemistry than by pushing single atoms around with a tiny tip, to get them to go where you want them to go. But there's no question that writing the IBM logo in atoms was something of a landmark in showing that technology, in the broadest sense of being able to manipulate matter, now reaches down to the smallest scales imaginable.

Philip Ball is a London-based science writer and prolific author.

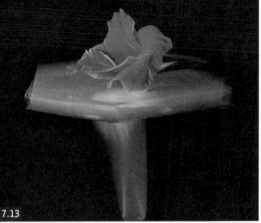

SUPERLIGHT SOLID. Aerogels were developed in the 1930s, but have only recently found practical applications in space. Particles shot into aerogel at high velocities in an experiment leave carrot-shaped track marks (7.12). The insulating properties of aerogel protect the flower from the flame (7.13). A block of aerogel weighing only 2 g can support a brick weighing 2.5 kg (7.14).

7.15 SUPERLIGHT STRUCTURE. It is the lightness of plastic films and molded structures that have helped innovative designs like the Eden Project in Cornwall, England, which houses more than 100,000 plants representing 5,000 species from many of the climatic zones of the world.

Superlight: Aerospace engineers and architects, respectively, benefit enormously from the efficiency and versatility of porous gels and flexible films. In the realm of the superlights, less is always more.

A block of aerogel as large as a human may weigh less than half a kilogram (or less than a pound), yet support the weight of a subcompact car (about 454 kilograms, or 1,000 pounds).

– The Jet Propulsion Laboratory (JPL) aerogel brochure

A material's lightness is important to consider when designing such things as mobile structures, portable appliances, and fuel-efficient vehicles. With handheld electronic devices and electric cars, superlight lithium batteries are used in place of heavy lead-acid batteries; lightweight materials are especially critical for electric vehicles, since the point with these is to conserve energy.

Carbon-fiber materials such as nylon and Kevlar, both light and strong, are commonly used for sporting equipment associated with speed, like car racing and cycling, and in the aerospace industry, where aerogel does most of its work today. Aerogel is as much as 99% air and is typically made out of silica, but it can be made out of a wide variety of materials, including carbon and polymers. According to *The Guinness Book of World Records*, the latest and lightest versions of aerogel weigh just 1.9 mg/cm^3 and are produced by the Lawrence Livermore National Laboratory in California. Although it is the lightest solid on Earth, aerogel is primarily used aboard spacecraft as a collection device for interstellar and cometary dust. Its pores and particles are smaller than the wavelength of light and it has low thermal and sound conductivity. Aerogel is the thermal insulation material of choice for the Warm Electronics Boxes (WEBs) on the 2003 Mars Exploration Rovers.

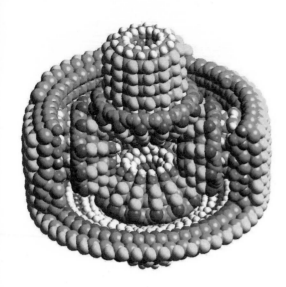

7.16 ATOMIC LOGO. The Scanning Tunneling Microscope made it possible for the IBM logo to be spelled out in atoms.

7.17 MOLECULAR GEAR. While a Research Fellow at the Institute for Molecular Manufacturing, K. Eric Drexler designed this molecular differential gear, a nanostructure to be built chiefly of hydrogen, carbon, silicon, nitrogen, phosphorus, oxygen, and sulfur atoms. It is analogous to a macroscopic machine, with rotating shafts and gears to drive motion. Drexler says, "Such structures are far beyond the state of the art of chemical synthesis today, but their design and modeling is becoming straightforward."

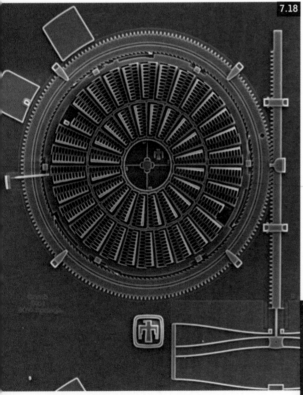

7.18 MICRO MACHINE. Sandia Labs has successfully fabricated Micro Electro Mechanical Systems (MEMS), micron-scale working machines.

7.19 NANOTUBES. In 1996, Sir Harold Kroto was jointly awarded the Nobel prize for chemistry with Richard Smalley and Robert Curl of Rice University, Texas, for the discovery of C_{60}, which led to the development of carbon nanotubes, shown below.

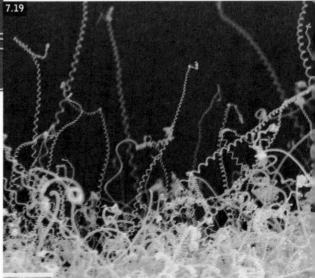

Supersmall: Through electron and atomic-force microscopes, physicists and chemists are looking to nature to build materials from the bottom up. The impact of this unprecedented development – nanotechnology – has yet to materialize on the macro scale.

I am not afraid to consider the final question as to whether, ultimately – in the great future – we can arrange the atoms the way we want; the very atoms, all the way down!
– Nobel Prize–winning physicist Richard Feynman (1918–1988)

Nanoscience is the study of systems with nanometer dimensions and the manifestation of Richard Feynman's big idea more than 40 years ago. According to George M. Whitesides, Mallinckrodt Professor of Chemistry at Harvard University, it is "a contender with genomics for changing the world."

Ever since IBM famously positioned 35 xenon atoms into the form of its corporate logo, it has been widely accepted that we can manipulate matter at the atomic scale. What hasn't been clear is where we go from here. While scientists develop this field and debate it hotly, terrifying notions of gray goo are seeping into the public psyche, thanks to scenarios put forth by Michael Crichton in *Prey* and to scientists like Bill Joy and K. Eric Drexler, who discussed nanotechnology publicly for the first time in the context of self-assembling nanorobots.

Chemists have a long history of designing structures from the "bottom up," and so their input is valuable here. In a recent point/counterpoint between Drexler and Rice University chemist Rick Smalley, Drexler compared his proposed (dry) assemblers to enzymes and ribosomes. Smalley: "Enzymes and ribosomes can only work in water. Please tell us about this new chemistry." Drexler has since backed down from his original nano-claims (from 1986) and has publicly admitted, "Runaway replicators, while theoretically possible according to the laws of physics, cannot be built with today's nanotechnology toolset."

The big question remains: Now that we can move atoms, what will we do with them? Perhaps mimicking biological, not mechanical, systems will lead us to our answer.

SELF-ASSEMBLING.
Cross-shaped self-assembled
structures (7.20) and scanning electron microscope
(SEM) image of polymer
micro-structures formed by
molding in capillaries (7.21).

7.20

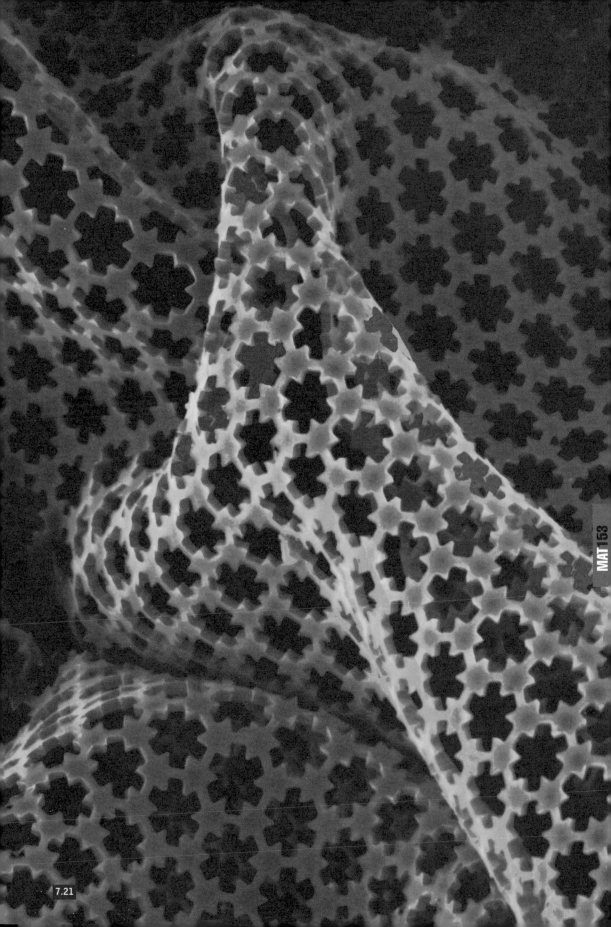

7.21

7.22 SELF-HEALING. An optical microscope image of Scott White's self-healing plastic. The microcapsules are colored red and the catalyst is black (the dark specs in the image). The healing agent has penetrated through the crack front, the solid red line across the center of the image.

7.23 FERROMAGNETIC. A ferromagnetic liquid is a fluid in which fine particles of iron, magnetite, or cobalt are suspended, typically in an oil. The fluid is attracted to magnetic force and can form into intricate patterns. Ferrofluids were invented by NASA as a way to control the flow of liquid fuels in space.

7.24 SELF-TIGHTENING. An image sequence of a thermoplastic shape-memory polymer developed by Andreas Lendlein shows the transition from the temporary shape of a straight rod to a self-tightening knot. It has use as a suture that ties itself during minimally invasive surgical procedures. The switching temperature of this polymer is 56°C. After heating up to 60°C, the recovery process takes 10 seconds.

Supersmart: Multidisciplinary teams of scientists are joining forces to design materials with built-in stimulus response. These smart materials can be customized with sensitivities to signals such as heat, light, impact, pulses of electric currents, and motion.

The material is the mechanism.

– Philip Ball, science writer and consulting editor at *Nature*

While chemists research the realm of the supersmall with the mission to develop enzyme-like tools to construct supersmart self-assembling materials, tissue engineers are building polymer scaffolds that support the growth of human organs and tissues.

Robert Langer works at the interface of biotechnology and materials science. He is the Kenneth J. Germeshausen Professor of Chemical and Biomedical Engineering at the Massachusetts Institute of Technology (MIT). He and Jay Vacanti, the John Homans Professor of Surgery at Harvard Medical School, initiated this burgeoning field of tissue engineering and have successfully synthesized new biodegradable polymer systems that have supported the growth of livers, cartilage (nose, ears), and nerves. The cells that are seeded on these structures are smart "natural" materials; they manage to recreate their respective tissue functions. In 1998, Langer and Professor Andreas Lendlein, of the University of Potsdam, founded mnemoScience, an MIT spin-off, specializing in the design of smart synthetic materials based on shape-memory polymer technology. German-based mnemoScience was recognized as the country's only Technology Pioneer in 2004, a designation given annually by the World Economic Forum.

At the University of Illinois at Urbana-Champaign, Scott White studies self-repairing plastics. He is an associate professor in aeronautical and astronautical engineering and takes his design inspiration from the rhinoceros horn. He and a multidisciplinary team of scientists designed a biomimetic polymer with embedded capsules full of "healing" liquid that, upon rupture, self-corrects cracks in plastic and fiberglass. According to White, this self-healing plastic can be used anywhere synthetic polymer is used now, from microchips to the wings on a full-size aircraft.

Janine Benyus
on biomimicry

With our increased capacity and sophisticated scientific tools, do we have the opportunity to align ourselves more closely with nature than ever before?
More powerful micro- and macroscopes are allowing us to see the inner workings of a dragonfly's wing, and to watch in color as a star is born. As biological knowledge doubles every few years, we have more information to inspire us – more evolved designs and strategies we can learn from. That's a great trend pulling us toward new kinds of innovation. At the same time, our tools are feeding back disturbing news: the double-glazing of the earth, the toxins that we're swimming in, the water shortages. As organisms, we're feeling biologically vulnerable again, but in this case it's not the saber tooth tiger that's threatening our lives. It's us. And that discomfort is providing the push toward innovation. We're realizing that we need to use our scientific savvy to change the way we've been living our lives.

Materials scientist Mehmet Sarikaya has said, "We are on the brink of a materials revolution that will be on par with the Iron Age and the Industrial Revolution. We are leaping forward into a new age of materials. Within the next century I think biomimetics will significantly alter the way in which we live." Describe his work in this context.
Mehmet studies a giant marine snail called abalone, whose mother-of-pearl interior, in addition to being incredibly beautiful and iridescent, is twice as tough as our high-tech ceramics, which we still make today by the heat, beat, and treat process. We heat kilns up to 4,200° F when we make ceramics. The abalone shell doesn't do this, of course. So Mehmet wondered if we could make materials at seawater temperature. The real breakthroughs now are coming in the area of self-assembling materials that mimic the way the abalone creates its puff-pastry architecture of hard mineral and soft polymer. Jeff Brinker, who's at the Sandia National Laboratories in Albuquerque, New Mexico, works in the area of self-assembling materials. This will be the way we make products in the future. It's coming.

What did you learn from the late biocomputing futurist Michael Conrad?
Michael Conrad felt that every cell in our body is a sophisticated computer, in the sense that it is taking in data all the time. But it computes with these self-assembling, three-dimensional molecules that jigsaw together. This is computing through shape, a completely different paradigm than computing with ones and zeroes on silicon. This is computing on carbon, also known as wet computing.

What sort of work is going on at Arizona State University in Tempe that could revolutionize our energy industry?
There's a group of 25 scientists or so who work at the Center for Early Events in Photosynthesis, who study the first few picoseconds, literally. They're mimicking a cell and doing chemistry with sunlight.

What does it mean to grow food like a prairie?
Wes Jackson runs a research organization called The Land Institute in Salina, Kansas. He realized that industrial agriculture is a failure because it grows miles and miles of annual plants in a monoculture. To plant annuals, you have to dig up the soil each year with tractor diesel, leading to massive erosion. To recoup fertility, you add fertilizer made from natural gas. The monoculture is an all-you-can-eat restaurant for plants, so you add pesticides, another fossil fuel product! It's ten kilocalories of oil for every kilocalorie of corn. He asked, "What if we were to learn from the prairie next door?" A prairie is a perennial polyculture, so for 27 years he's been trying to perennialize our major crops and plant them in mixtures. His scientific staff now is world-class. But, of course, he could use lots more scientists working on this.

Let's play a word association game. Hummingbirds.

Poster child for sustainability. In the process of getting the fuel they need through nectar, hummingbirds manage to pollinate flowers, which ensures that there will be fuel next year for them or for other hummingbirds. That's the kind of system I would like us to have at our gas stations.

Chimpanzees.

Chimps are tremendous teachers in terms of what we should be eating. They are masters of smart eating, and now we realize that they self-medicate. They're able to choose certain plants in particular bushes and prepare them in particular ways to heal themselves. Howler monkeys pick plants that will actually bring on fertility during the reproductive stage, or delay it. So they seem to have not just self-medication going on but also birth control.

Purple bacteria.

A very ancient kind of bacteria that's being studied because they photosynthesize. It also has a pump

We need to find the recipe and cook it in our own kitchen, and leave the organisms alone. That to me is true biomimicry.

within it that would make hydraulic engineers jealous. When lit by the sun, it pumps ions through its membrane, upstream, against gradient.

Rhinoceros horns.

Rhinos dig with their horns and use them as swords to spar with. If they get a crack in the horn, it's a problem. Surprisingly, we found that they don't seem to get cracks in the horn, and if they do, the cracks seem to heal up. When there's a crack, the material around the crack disassembles, pours into the crack, and then reassembles. We have no idea how this happens because there are no living cells in the horn. It's made entirely of dead hair! We're looking at it as a model for self-healing structures.

In my materials research, I've come across the work of Scott White at the University of Illinois at Urbana-Champaign. He's working with self-healing plastic.

Yes. There are self-healing plastics and self-healing concretes. We basically put [in concrete] little capsules of an epoxy, of a gluelike material, that when the concrete flexes will break and pour into the crack. It's an exciting but crude approximation of the amazing process that goes on in rhino horns.

Coral reefs.

Well, besides the fact that they're bleaching worldwide, which is a very big problem, coral reefs are a tremendous model of what's called a Type 3 ecosystem. A Type 1 ecosystem is an ecosystem that's temporary, like a carpet of weeds in an open field. Berry bushes move in and then eventually it's a forest. A mature forest – like a coral reef – is a Type 3 ecosystem. These are complex communities, an incredible society of organisms that are in deep symbiosis with one another. Business managers are looking at these Type 3 systems, believe it or not, as a model for a new way to organize our whole economy so that it's much more interconnected, interdependent, and less of the competitive model that has a lot of inherent waste in it. More and more, it's about the closed-loop, tightly woven food web, where nothing is wasted; there's always somebody there to catch that next little particle of food and use it.

Have we been successful in mimicking spider silk yet?

This incredible material is five times stronger than steel. In *Biomimicry* I wrote about researchers [Fritz Vollrath et al.] who were very keen to figure out how to make this fiber the way a spider does – in water and at room temperature – without petroleum, without heating it up, and without sulfuric acid. Since then, Montreal-based Nexia Biotechnologies has decided to take the gene out of the spider and put it into a goat and then have the goat milk the protein, which turns into the fiber. The transgenic goat milks the protein from its mammary gland. And then the goat gets cloned. Gene transfer from spiders to goats doesn't happen in the natural world! Besides, this is just domestication with a twist. To truly emulate would be to find the recipe, find the way to manufacture in water at room temperature, like Jeff Brinker did with abalone. We need to find the recipe and cook it in our own kitchen, and leave the organisms alone. That to me is true biomimicry.

Janine Benyus is the author of Biomimicry: Innovations Inspired by Nature.

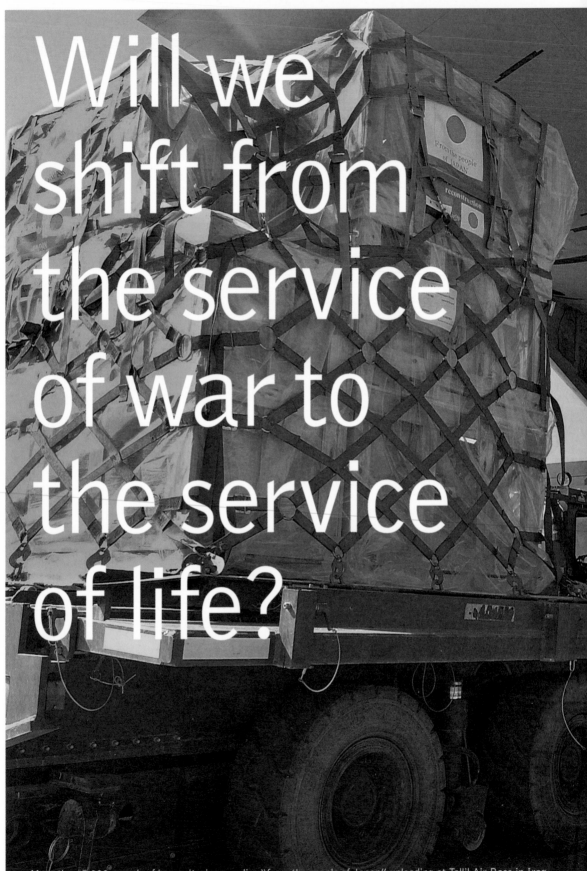

Will we shift from the service of war to the service of life?

More than 5,000 pounds of humanitarian supplies "from the people of Japan" unloading at Tallil Air Base in Iraq.

The design of defensive and offensive technologies, a practice centered on raw efficiency, has generated the twentieth century's dominant cultural mode. Innovations developed by the military have migrated to almost every design practice – from material development to command and control, to robotics and communication – providing exponential impact in the civilian sector. We are living in a "war machine," as renowned urbanist and military theorist Paul Virilio sees it. Can we reimagine our use of military-derived technological power?

MILITARY ECONOMIES

8.01 DESIGNER SPIN-ON. The Eames Studio, one of the shining stars of American design, was founded on the military project. Using their revolutionary molded plywood technology in early 1942, Charles and Ray Eames took on the challenge of redesigning the standard metal leg splint for soldiers wounded at war. After long hours of uncomfortable plaster mold experiments on Charles and trimming and slotting by hand, the Eameses presented the resultant design to the U.S. Navy in San Diego. A few iterations later, in November 1942, the navy approved of the revised prototype and placed its first order for 5,000 splints.

One of the most challenging subjects to come to terms with in the writing of this book was the military economy. At times we considered dropping it altogether simply because it runs so fundamentally counter to British historian Arnold Toynbee's ambition – and the thesis of our project. But, at the same time, to omit it would be to escape from the responsibility of discourse.

We discovered a real contradiction in our research: For better and worse, the military project continues to be one of the most powerful engines of technological innovation and design. From the microwave oven to space exploration, from civilian aviation to the Internet, the military (and in particular the U.S. military) has fed the process of design.

We cannot consider our global future without considering the impact of the military economy, both its capability for madness and its capacity for action on a global scale.

Gwynne Dyer
on conventional war

Is it true that the introduction of precision weapons has reduced the number of soldier deaths in war?
It's true that the number of American soldier deaths has radically gone down. The number of soldiers on the other side has not. The whole thrust of American technological development in weaponry – conventional ground and air force weaponry – for at least fifty years, undisputedly, has been to spend virtually any amount of money to reduce American casualties. Americans actually now believe you can have wars with no casualties on their side – but, remember, this only works against technologically inferior opponents. America hasn't fought any technologically equal opponents since 1945. The U.S. has had a long, easy run of it. The implications of all this technology would be very different if both sides were on a level playing field.

Tell me about these so-called intelligent machines.
The machines we're talking about – these precision-guided weapons – are not intelligent in any kind of meaningful sense. They're guided or self-guided. The jargon is "one-shot kill" capability. That is to say, in the First World War, you fired at least ten thousand bullets for every casualty. In the Second World War, you fired fifty to a hundred artillery rounds for every casualty you inflicted. Now, you fire one weapon and inflict a casualty – the casualty may well be a tank or

an airplane, not just a person. The goal of all this stuff is to reduce the amount of weaponry you have to expend and kill the opponents. Of course, the weapons are so expensive that you'd better kill somebody with every one you fire; each round can cost anywhere from ten thousand to half a million dollars.

What comes to mind when you hear the term "future warrior"?
First of all, I'm not really keen on future warriors. I don't think we need them. There is this fantasy that now that we've solved the military problem and, with these wonderful weapons, we can win a war without casualties. Yeah, sure you can, if you're fighting three-generations-old technology or people who haven't got weapons at all. But even then, you'd win the first battle and, afterwards, if you actually want to stay around and occupy the territory, you're back in the old grim world where a roadside bomb is just as good as an Apache helicopter. High technology is great for defeating conventional military technology that is two or three generations out of date, but it doesn't solve the political problem you came to solve. When and if the opposition drops down to guerrilla warfare or resists with terrorism, for example, high-tech weapons become useless.

Why are we still spending so much money on weapons of mass destruction?
You're familiar with the concept of cultural lag? I think it's pretty relevant here! I've been hanging around the military all my life and they're smart people, but they are definitely wedded to their profession and the kind of wars that their profession has always fought. Conventional warfare between nuclear-armed countries has become redundant and ridiculous, however.

What ever happened to Lester B. Pearson's notion of peacekeeping?
Peacekeeping in the full-blooded, very institution-alized Pearson sense of the word – he got the Nobel Prize for inventing it – was actually designed as a buffer zone during the Cold War. The idea was that, if you have conflicts you want to put on ice and you don't want to have a Soviet-Western confrontation, then you freeze it with peacekeeping troops. This was what was done in Egypt in 1956, when the British and the French conspired with Israel to invade Egypt, and it got way out of control. They wondered, "How do we get out of this mess?" The answer was to put peacekeeping troops in there. And

Canadians went. That really was Pearson's contribution. After the Cold War, this concept was expanded into a more interventionist, humanitarian kind of operation. So nothing has happened to Pearson's notion of peacekeeping. It continues to serve humanitarian functions today, as well as geo-political functions. A lot of it has now been sub-contracted to NATO.

Is the European Union (EU) a good model for a peaceful future?

I think it's an excellent model for a peaceful future. But the thing that makes the Europeans so reasonable now is that they were so astoundingly unreasonable for a hundred years, and nearly exterminated themselves. Every city, basically, was bombed flat in Europe by 1945. A lesson has been learned, which has not been, mercifully, administered to the United States. So the Americans have a very different and, you might say, more naive view of what is achievable by way of war.

Are you referring to President George W. Bush's Pax Americana?

Oh, yes. George W. Bush and the folks around him, the neo-cons, do have a very ambitious project. Pax Americana, based on the two-thousand-year-old Pax Romana, is essentially arrogant, but well meant. If you go in and look at the details, there is the intention to make a profit on this too, but the notion that we do what Rome did and extend our authority over

High technology is great for defeating conventional military technology that is two or three generations out of date, but it doesn't solve the political problem you came to solve.

the planet because we have the power – and, look, you can bring peace to the world by this – is at least partly well meant.

What is the fate of the UN after the most recent events in Iraq?

Pax Americana requires the destruction of the UN. That is the long and the short of it because the principles are opposed. The UN's founding principle is that war is now illegal. You cannot attack another country. The Pax Americana project's core rule is that the United States will decide what countries are dangerous and will take them down, unilaterally, without consultation, and without help, if necessary. It will be judge, jury, and executioner. You can have one or the other, but you can't have both. The invasion of Iraq was illegal by UN rules. It was compulsory under Pax Americana.

But Pax Americana is going to fail. It's as certain to fail as the Islamist project is going to fail. These projects both come very much from the margins of their respective societies. Pax Americana is not a mass movement. Invading Iraq was the launch vehicle for Pax Americana, in a sense. That's what Iraq was really about, behind the facade about weapons of mass destruction or links with terrorists or all that other nonsense, which has now been blown away.

What are the alternatives to war?

The institution of war is older than civilization but, in a recognizable sense, it's probably six thousand years old. It made a certain amount of sense when the only real source of wealth was land – of which there is a limited supply. But land is not really a source of any nation's wealth anymore. The whole economy has moved beyond that. Wealth now derives mostly from industry and innovation and intellectual property, which can only be destroyed by war.

There will always be disputes, as long as there are human beings, but there are very few disputes that you would rationally choose to solve by war anymore. Over time, new institutions and relationships will evolve that undercut the whole nation-state model, which is the frame-work within which war occurs. We need the UN right now because we are still in the nation-state model, and it's an attempt to keep us from blowing ourselves to kingdom come before evolution carries us further on. I cannot tell you what in 2073 the prevailing model is going to be, but there will certainly be large traces of the present model, in conjunction with new elements and relationships, as the old, exclusive ideas of national identity slowly erode and we become ever more interconnected on a global scale.

Gwynne Dyer is a London–based journalist and military historian.

MIL163

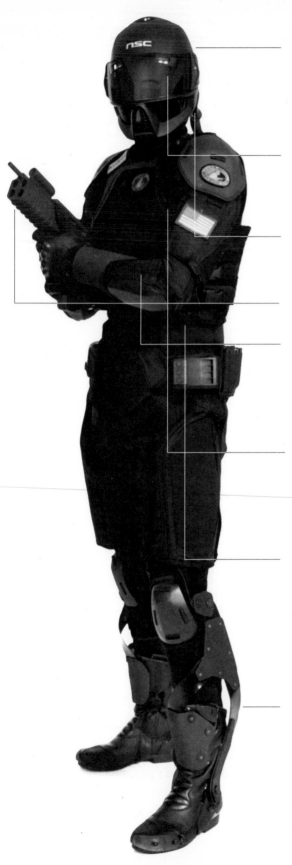

8.02 FUTURE WARRIOR? The U.S. Army's concept figure, "Future Warrior", wears a helmet equipped with a 3-D audio and visual sensor suite, which restores natural hearing lost in an encapsulated space and enhances long-range hearing. A small halo on the helmet represents a tracking system for friendly and enemy forces.

The blue-tinted visor signifies agile eye protection against tunable lasers and sensors for thermal and image intensification, which are small enough to be projected onto the visor. Cameras enhance vision from the sides and behind.

A respirator tube at the back of the helmet connects to a low-profile air purifier that forces cool air into the helmet for comfort and visor defogging, which protects against chemical and biological agents.

The weapon is a fire-and-forget system, using soft-launch seeking missiles.

A flexible display on the forearm of Future Warrior glows when switched on and draws attention to the simulated touch-screen keypad for information input and output, and for tasks such as navigation, physiological status monitoring, and command communication; the display is connected into a compact computer worn on an armored belt around the waist.

Near the top of the torso, front and back, are quarter-sized buttons built into the fabric, depicting a nanostructure sensor array to detect weapons of mass destruction, friendly or enemy lasers, or even weather; the sensors could trigger a response in the uniform to open or close the fibers depending on temperature or precipitation.

Research by the Massachusetts Institute of Technology on nanomuscles for advanced arm and torso strength may be linked to the exoskeleton to give Future Warrior potentially superhuman ability to move or carry. Along the black stretch fabric are custom-fitted plastics and foams that take the place of liquid body armor that will instantly solidify when struck. Nanotechnology will give multifunctionality to materials, such as the abilities to transport power and data, fend off chemical and biological agent attacks, self-decontaminate, and become waterproof.

Protruding, interconnecting black pieces of plastic on the legs represent a lower-body exoskeleton, which will connect through the boots up to the waist and enable the wearer to carry up to 200 pounds.

All this technological wizardry: amazing, but crazy. What are we communicating, if not total domination? Technological solutions to what are social and cultural conflicts miss the point entirely.

Consider the human above the technology: If war is inevitable and one death is too many, is it possible to wage war without killing anyone?

Before one begins to be able to recognize the human being in the other – whether it's an enemy or just somebody else – I think the main challenge one has is to recognize the human being in oneself.

– Sari Nusseibeh, president of Al-Quds University, Jerusalem

There's a lot of talk these days about the "future warrior" and, depending on whose home turf you're on, he is either high-tech or rogue, military or paramilitary. On the one hand, said warrior benefits from digital tools, imaging equipment, and high-functioning materials. On the other, he makes do with found materials, hand-me-down weapons, and inventiveness. In both scenarios, he plays to win.

If, as military historian Gwynne Dyer suggests, conventional reasons for going to war (land, primarily) are fading away, then why not entertain the possibility of designing a new model of warfare, without the bells, whistles, and body bags?

In an age of information warfare, where the distinctions between virtual and real are fuzzy at best, why are we imagining a future warrior who's armed with tools for killing anyway? Why not envision our soldier of the future stepping away from the killing fields towards a war zone that supports nonviolence? Arm him with cultural history, diplomacy, language skills, courage, charisma, and the ability to take an open-minded view of enduring conflicts. Help him recognize the human being in another.

This is unabashedly utopian, but – considering the alternative, a soldier decked out in "Terminator" gear, whose primary training takes place in computer-simulated worst-case scenarios – it's more closely aligned to Arnold Toynbee's practical objective of considering the welfare of the entire human race.

When presented with the choice, which future warrior will we choose?

James H. Korris
on virtual war

compelling simulation and learning systems for the Army.

What does Hollywood bring to the mix?
Hollywood brings story and character to simulations. In fact, one of the first things we discovered in our working together was that people in the military tend to view a story as a straightforward series of events. Hollywood views a story as a series of linked events with deliberate complications and predicaments designed to build to a dramatic climax and evoke a particular emotional response from the viewer. In a simulation, the reasoning goes, if people are emotionally engaged, they will pay more attention and learn better.

How are video games training soldiers today?
Simulating real world problems better prepares soldiers for the human side of war and conflict and the kind of decision-making soldiers are faced with in urban environments. Our aim is to create an immersive learning experience with video games. Our challenge is to push the technology well beyond where it is.

Why did the U.S. military join forces with Hollywood and the electronic games business?
For years, training simulations were, almost exclusively, the domain of the U.S. military. They conducted live simulations, used war games, and employed virtual simulations like the Close Combat Tactical Trainer and a number of aviation trainers. But these simulators were very specialized, tended to run on proprietary equipment (sometimes mainframes) and, with specialized displays and interfaces, could cost millions. By the mid-1990s, the military found that in simulation technology they had been, in some ways, leapfrogged by the entertainment software and computer game industries. The investment in R&D [research and development] by the interactive entertainment business exceeds that of the Army in graphics hardware, for example. The military tends to be a very young group, so you had about 25% of kids joining the military who considered themselves serious gamers. Probably all of them had played computer games at some point and many continued to play games. There was a real disconnect between the quality of stuff they worked on in the military and what they were used to playing on their X-box. So the military proposed to the University of Southern California that their computer research scientists, professionals from the Hollywood-based film community and the electronic games community collaborate on developing enabling technologies for immersive virtual reality and realistic and

What are the major areas of research at the Institute for Creative Technology (ICT)?
We create immersive virtual worlds, using graphics and sound, with virtual autonomous agents who support face-to-face interaction with people in realistic environments. We work with Artificial Intelligence (AI) in the development of virtual humans who can interact with real humans; we also work with emotional modeling, character development, natural language recognition, and speech.

8.03 Screen shot from the ICT's Mission Rehearsal Exercise System virtual training environment.

Character simulations have been fairly rudimentary in the past. What challenges do you face when trying to create realistic characters?

That's really our biggest challenge. Virtual humans represent an omnibus of research areas. In order to have a meaningful interchange with a computer-generated person, there are a lot of moving parts that need to be present. One is a natural language understanding; the ability to comprehend human speech is very daunting. If you parse and try interpreting most human speech, it makes no sense at all. So natural language understanding tries to make it possible for a human to try to communicate with a machine in a conversational, natural way. There's also speech synthesis. Once you've decided how a machine should respond, how does it respond in an audible way that sounds natural?

How about the emotional model? If someone crashes his or her shopping cart into yours in the supermarket, you could ask, "What are you doing?" The person might respond with, "Hey, watch where you're going!" If you asked that question of a machine it would respond literally: "I smashed into your shopping cart." The quality of defensiveness or shyness or aggression isn't there. So, you have to work out a context and, from this, construe a likely emotional posture before you can have meaningful discourse.

There's also a significant body of research called cognitive modeling that tries to deconstruct the way people imagine they think, as opposed to the way they actually think; this is about what's going on inside your head. To that we add facial animation. When we talk to people, we obviously get a lot of cues from whether they're frowning or smiling or looking us in the eye or looking away. We infer things from that. So facial animation and gaze direction are important. The other thing is whole body animation. Beyond this entity-level speech exchange, a virtual human has to eventually task-organize – work with other virtual humans or with humans – task-plan, and then go do something. We can't have him walking into walls. We have to build in obstacle avoidance. The simple act of standing up and leaving the room is a big deal if you're a virtual human.

You recently won an award from the computer games community for one of your first productions. What kind of feedback are you getting from the military?

Right, we won two E3 Critics Awards for Full Spectrum Warrior. Pretty much everyone who has played this game loves it. The critics recognized it

as a great entertainment game, but what they didn't know is that it was developed as a military trainer. One of our researchers played the game for two days while he was demonstrating it at a military conference. In that time he only successfully completed one mission! It's a hard game. It's all about one's ability to make tough decisions under pressure of time with inadequate information. The Army Research Institute evaluated another game, Full Spectrum Command, for which we scored very well on engagement, involvement, and on the relevancy of game tasks. We produced a board game version of it first and a PC version, which built on what we learned from making the board game. As a corollary to that project, we're now making two training aids for the Singapore Ministry of Defense, via a technology partnering agreement between the U.S. government and Singapore.

Saving the credits for the end, who are some of the movie people and games people on the ICT team?

We've been fortunate so far in attracting incredible people, both from the games side of things and the entertainment side. With respect to film entertainment, Randal Kleiser (*Grease, Blue Lagoon*) and John Milius (*Apocalypse Now, Clear and Present Danger*) got involved early on and remain involved in various projects. Dick Wolf (*Law & Order*) was involved in a couple of projects. We also work with Christopher Crowe, who's done a lot of television and motion picture work (including the recent *Homeland Security*) and wrote the screenplay for *The Last of the Mohicans*. Also: Joe Zito, who's known for his action movies; Stephen de Souza, who wrote screenplays for the *Die Hard* series and *48 Hours*; David Ayer, who's best known for having written *Training Day*, *U571*, *SWAT*, and *The Fast and the Furious*. Ron Cobb, a production designer whose work goes back to the canteen scene in *Star Wars*, has been designing concepts for vehicle platforms interface for the military. On the games side, we usually work with companies, although individual efforts have come from J.C. Hertz and Mark Perenski. We've worked with Sony Image Works, who did the initial art assets for Full Spectrum Warrior; Pandemic Studios, our development partner on Full Spectrum Warrior; Quicksilver Software Inc., who worked with us on Full Spectrum Command; and Legless Productions.

James H. Korris is creative director of the Institute for Creative Technologies at the University of Southern California in Los Angeles.

Arthur Kroker
on cyberwar

What does it mean that war is now mediated through technology?
Today not only the act of war itself, but the perception of war is a technological event. In a significant way, there are always two theaters of war: actual battlefields with real casualties and immense suffering, and hyperreal battlefields where the ultimate objective of the war machine is to conquer public opinion and manipulate human imagination. Particularly since 9/11 and the prosecution of the so-called "war on terrorism," we live in a media environment which is aimed at the total mobilization of the population for warfare. For example, in the American "homeland," mobilization of the population is psychologically conditioned by an image matrix, fostering deep feelings of fear and insecurity. This is reinforced daily by the mass media operating as a repetition machine: repeating, that is, the message of the threatening "terrorist Other." For those living in the increasingly armed bunkers of North America and Europe, we don't experience wars in any way except through the psychological control of perception through mass media, particularly television. The delivery of weapons – themselves intensely sophisticated forms of technology – are part of the same system. So tech-mediated war is the total mobilization for warfare with us as its primary subjects and targets.

What is the effect of our seeing from the bomb's eye view?
Perhaps human vision itself has now been literally harvested by the war machine. When we see the unfolding world from the bomb's-eye view, this would mean that what we traditionally have meant by human perception – vision, insight, ethical judgment, discriminating between reality and illusion – has been effectively shut down, almost surgically replaced by the virtual vision machine of the militarized imagination. We are suddenly rendered vulnerable to the new virtual myths about the supposedly hygienic character of post-human warfare. For instance, the spectacle of the bomb's-eye view supports the illusion of war as being about so-called "smart" bombs, which are hyped as controllable in their targeting trajectories, with few civilian casualties. The audience becomes a spectator of this act, but it's a complete fabrication. Only long after the first Iraq war was it revealed that many of the cruise missile shots, which were supposed to be precise in their "target acquisitions," may have been staged video shots. The reality of that war had to do with massive bombing raids and antipersonnel cluster weapons, all of which were deliberately aimed at civilian populations.

How do you respond to the argument that precision weapons are the reason behind lowered death rates in war since WWII?
This is a very complex question. Industrial wars such as World War II have a necessary accident: high casualty rates both among the civilian population and mass armies of soldiers. In the post-historical time of assured nuclear destruction, mass conflict was avoided but the planet witnessed a contagious growth of local political wars, many of which were directly linked to the struggle for global supremacy on the part of the bipolar powers of America and Russia. In the unipolar world of the American empire, power is maintained by military strategy aimed at "full-spectrum dominance" by an increasingly cyberneticized military apparatus. The empire fights for total sovereignty over both space and time. It seeks to virtualize warfare, reducing the unpredictable nature of urban war to the cybernetic certainties of precision weapons, cruise missiles, and laser-guided bombs. However, it is the fate of all otherworldly illusions to finally succumb to earthly realities. Consider the two Gulf Wars, which may have been state-managed in the language of precision weapons and low civilian casualty figures, but were typified by antipersonnel cluster bombs aimed at terrorizing the Iraqi populations. Mass media do not discuss Iraqi civilian casualties, since it is in the nature of empires to make the humanity of scapegoated populations literally disappear. Perhaps we should keep in mind that the ultimate casualty in the new era of micro-

warfare is the death of political hope and an ethics of reciprocity.

When you think about all the technological innovation that has come out of the military, is there room for celebration in the civilian sector at all?
No [laughs], I don't think so. I really don't think so. One of the agitprop propaganda tactics of the military in the space program, among others, is to say there's this incredible spillover of civilian applications for military technology, but that's silly because the base of military expenditure for the most part is what Steve Kurtz from the Critical Art Ensemble has called "useless technology." That is, military programs serve a very traditional anthropological function in society. The military is a site of sumptuary expenditure: It produces weaponry that are all about sacrificial

In the unipolar world of the American empire, power is maintained by military strategy aimed at "full spectrum dominance" by an increasingly cyberneticized military apparatus.

violence, either sacrificing the human race as a whole as the end game of nuclear assured destruction or sacrificing chosen scapegoats as a way of appeasing the psychological anxieties and appetite for destruction of society. If weaponry is not used, then it is fated for technological oblivion: Military culture today possesses, in fact, an enormous archive of outdated weaponry that can neither be effectively used on the battlefield nor successfully recycled. Sometimes sumptuary military innovations such as spent uranium are recycled in the form of armor-piercing artillery shells that will be discarded on foreign battlefields or, in the case of the present Iraqi war, sumptuary military expenditures take the form of a spectacular scene of sacrificial violence (the night bombing of Baghdad), which was probably aimed at the real media target of the campaign of "shock and awe" – namely, the domestic population of the American homeland. Could it be that the real crossover of military innovations is in the area of the militarization of the media imagination? If so, we are then speaking about the innovative design of moral perversity as the true spearhead of the empire of technology.

Martin Heidegger went to his death convinced that the question of technology was coeval with the ascendancy of the will to annihilation. Please explain.
For Heidegger, the apocalypse had already happened. With the dropping of atomic weapons on Japan at the end of WWII, human history itself was finally brought under technological control. He said that humanity had been reduced to "standing reserve," waiting to be harvested by a nihilistic technological apparatus. From Heidegger's perspective, this period of world history is immensely threatening and convulsive.

What do you think Marshall McLuhan would say today if he were alive?
McLuhan might say that in moving so quickly from hot wars to cold wars and then to viral warfare, under the sign of biotechnology, warfare itself has become its own virtual simulacrum: a self-contained, self-promoting, self-transforming virtual phenomenon stretching from the hard physics of dropping bombs and abusive physical violence to the psychological colonization of human perception. In this case, war is the skin worn by an empire that knows only the language of violence. For McLuhan, the empire of technology performs "psychic surgery" on the global village by means of the war on terrorism. While often viewed as a technotopian, McLuhan's perspective on technology and its possible applications was deeply pessimistic.

If you could give advice on the topic of developing filters to see through the military-mediated reality, what would it be?
I would restate the advice given by [German philosopher] Karl Jaspers in *Man in the Modern Age*. He said, "In times of radical crisis, everything depends on the individual who says no, who, acting out of the courageous impulses of human solitude, refuses to assent to a power that would be totalizing." Critical ethical judgment on the destiny of individual freedom is the very best filter of perception. As Albert Camus said in his wonderful book, *The Rebel*, "I rebel, therefore we exist."

Arthur Kroker is a cultural theorist and Canada Research Chair in Technology, Culture and Theory at the University of Victoria, Canada.

INF IMA

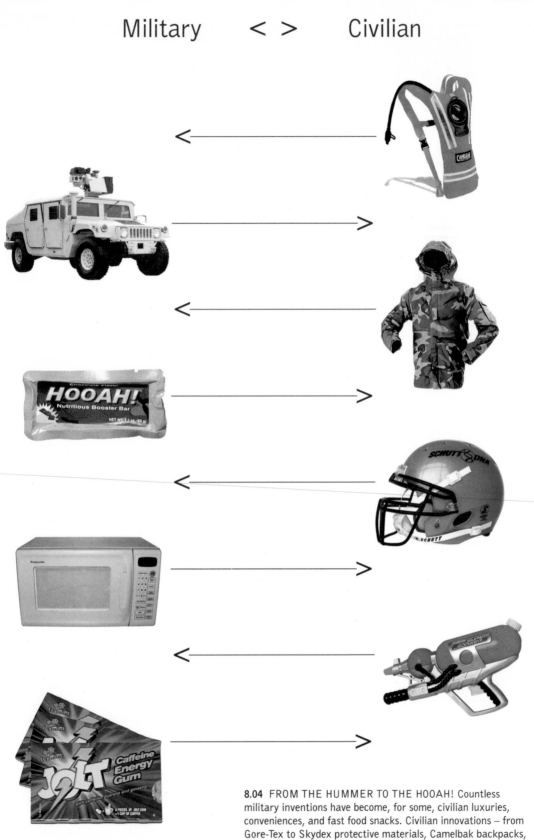

Military < > Civilian

8.04 FROM THE HUMMER TO THE HOOAH! Countless military inventions have become, for some, civilian luxuries, conveniences, and fast food snacks. Civilian innovations – from Gore-Tex to Skydex protective materials, Camelbak backpacks, and the Super Soaker toy gun (which inspired the M16 firearm) – have found their way into military culture.

Spin-ons and spin-offs: This is not about the transfer of technology. It's a dynamic intermingling of transferred processes: command-and-control and hierarchy mixed with collaboration and innovation. Who's influencing whom? And who's benefiting?

In a garrison society, it is difficult to differentiate between what is civilian and what is military. The functions seem to intertwine, overlap.

– Vernon Dibble, sociologist and author of "A Garrison Society"

Technological enhancements to conventional weaponry may be redundant in a world with nuclear arms, but there's no denying that the civilian sector has absorbed military-derived innovation.

We are all part of military culture, at times of war and peace. Whether we know it or not, we incriminate ourselves every time we use technological innovations known as "spin-offs," which have arisen from military-sponsored research and later get adopted by civilian society. Since Napoleon, the long line of spin-offs have included canned foods, plastics, microwave technology, radar, lasers, the Internet, night vision, jet engines, cell phones, GPS systems, Gore-Tex, frozen foods, and power bars.

British economist Mary Kaldor noted in *The Baroque Arsenal*, however, that most military research conducted post-WWII was for devices too specialized and sophisticated to have any application in the civilian marketplace, which thrives on cheap, dependable, mass-produced goods.

Since the second Gulf War, the tide has turned: the military has begun to make use of goods produced in the civilian sector. Aptly called "spin-ons," technological tools developed by the civilian sector that now benefit the military include off-the-shelf commercial information technology (computer workstations, laptop computers, database software, networks), specialized graphic processors developed by Hollywood and the video game industry (for simulators and communications and surveillance gear), and email, which allows for "open-source intelligence" and the critical debate around what for too long has been stamped "Top Secret."

Bruce Sterling on the next 50 years

In *Tomorrow Now*, you travel through Shakespeare's "seven stages of man" – from *As You Like It* – as a way of navigating the next fifty years. Why?
Tomorrow Now is a book about nearly everything. But you can't simply write a book about every aspect of the future because it's like writing a book about every aspect of the present. So the framework I decided to use was the human body. This book is very body-centric and I try to make human flesh and a human sensorium into a sort of key that opens the future. That's our real encounter with the future; it's not these abstract notions of drivers or changes, but the fact that time flows through your body and you can't really live unless you're moving into the future at the rate of one second per second.

You say genetic engineering is in its infancy. How is this new baby, so to speak, shaking up our global household?
Genetic engineering is barely getting anywhere and it's already subject to a great deal of controversy. If you look at what genetic engineering really does, as opposed to the things that it gets headlines for, it seems likely that it's going to get as close to the DNA as it can and as far away from the products of DNA as it can. The sorts of images we have of genetic engineering are the Frankenstein baby, Dolly the clone, and weird, monstrous, animals. When you look at what DNA is good for, it doesn't make sense to put it into humans or animals. It takes too long. DNA moves fast and the best way to carry it is in a micro-organism. Let's say you decide to create a super-baby next month using whatever techniques you found in [pioneering genomics researcher] Craig Venter's DNA lab. By the time this super-baby is an adult in the year 2024, there will have been another 20 years of further advancement in the field. So why would you make a baby with today's technology knowing it's going to be twenty years out of date when it's grown up and can actually vote?

Moving on to "The Soldier" now, can you explain the stylistic differences between the military and the paramilitary?
It's taking people a surprising amount of time to get their head around the idea that the bipolar world of communist/capitalist confrontation is over and we now have a confrontation between New World Order and New World Disorder. In other words, people don't get it that lawless narco-terrorism actually makes a lot of money and is a newfangled kind of mountain banditry. I mean, terrorism is not terror. When you're talking about a war on terror, that's like a war on technique. It's like having a war on Blitzkrieg, when the Germans would suddenly come over the Belgian border in tanks. What we really have is a serious disorder problem. We've got breakdowns in the Westphalian nation-state system because governments just can't control these huge living streams of illicit revenue from narcotics, arms smuggling, human smuggling, and so forth. It's not too hard for the U.S. to bomb anything from orbit. We can send over a B1 bomber from the heartland of the U.S. and have it literally circum-navigate the planet and drop munitions with absolute precision anywhere. But it turns out to be extremely difficult for the U.S. to walk block to block with military police trying to enforce order on people who really don't want any aliens around. So if you're in Serbia you're in big trouble – sort of – if the U.S. decides to take it upon itself to smash your government. They can smash all the government buildings and knock down the bridges and telecom centers, and so you have a hard time getting around. But if you're Somalia and there's nothing left to smash, the U.S. has got a problem.

The secret of the struggle between the New World Order and the New World Disorder is that they feed on one another. It's our own appetite for destruction that underwrites this warlord activity.

Do you see us rising out of a constant state of global warfare?

If I had my druthers between two paranoid superpowers with vast numbers of missiles pointing at one another prepared for a nuclear Armageddon at the drop of a hat, and the warlord stuff, it is absolutely a much better military scenario. You know, only 3,000 people died in [the 9/11 events]. Compare that to WWI or WWII, where it was nothing to have that many people die in a morning, month after month, year after year. Trench warfare of WWI was enormous mechanized slaughter. September 11 was bad, let's not kid around, but it was about as bad as Srebrenica. In the longer term, how many people could even spell Srebrenica? I mean, the political implications of it might be horrible, but we're doing that to ourselves. There's really not that much reason to flinch over the loss of two buildings in a large city when you compare it to what happened to Moscow or Stalingrad or the

When you're talking about a war on terror, that's like a war on technique. It's like having a war on Blitzkrieg, when the Germans would suddenly come over the Belgian border in tanks. What we really have is a serious disorder problem.

siege of Berlin or the occupation of Paris. Those were bad scenes! A spectacular terrorist theater event like 9/11 is, quite frankly, mostly media hype. That's what it was for. It was there to look spectacular.

In the context of recent events and Turkey's longstanding desire to join the EU, what do you predict?
The whole Moslem world is in really deep trouble in the twenty-first century; they're just having a real hard time getting a grip. You can see it from satellite photos. Compare a picture of India at night to Pakistan at night, and you can see who took the baton and ran with it. But if there's a solution for the problems of the Moslem world, I think it'll probably come out of a place like Turkey. And it may not look like anything we expect. But they're a surprisingly resilient and

inventive and really cheerful people and I have high hopes for them. I've spent a lot of time looking at what they're doing and I think they're great trendsetters.

You say oil is the most dangerous contraband in the world. How will things change as we work toward solving our energy challenge and grow out of our oil dependency?
This remains to be seen. Oil people are running the United States right now and they don't particularly want to grow out of this energy challenge. They make noises about a hydrogen economy, but if you look at the fine print they want to make hydrogen from coal. The fossil fuel industry is the biggest industry in the world and it really doesn't want to give up its revenue stream anymore than, say, the recording industry of America wants to give up its revenue stream, just because people have invented a new way to copy music.

The difficulty with oil is that it's fungible. You can dig it out of the ground and you can transport it. Practically everywhere there's oil there is massive human suffering now. The curse of oil is severe. Venezuela has problems. Nigeria has problems. Saudi Arabia, Central Asia, Iran, Iraq. Pretty much anywhere there's a wellhead there's a bad scene economically and politically. The Chechnyian war would have been over a long time ago if it weren't for the oil. The Iraqi war would not have lasted as long as it had if the U.S. wasn't paying for oil with one hand while it blew stuff up with the other. People have been mining this stuff for a long time and the process of corruption has become very intense.

When you look back on what you wrote about the next fifty years, are you optimistic?
I think the best attitude for a serious futurist to have is not pessimism or optimism, but just a deep sense of engagement. It has to mean something to you. You have to find aspects of it that can really compel you. And you shouldn't get hung up on whether it's "good" or "bad" because those qualities can change their coloration quite rapidly as time continues to pass.

Bruce Sterling is the author of Tomorrow Now: Envisioning the Next 50 Years.

Seymour Melman on the conversion project

When did you become passionate about the conversion project, from a military to civilian economy?
Around 1980 or so, a number of colleagues and I decided that the militarization of the economy was going full tilt, and that it was going to create very great difficulties for people finishing with military service. People who were spending their lives in military production were in for a very big letdown with the end of the war in Vietnam. And so, a group of us decided that we ought to form a National Commission for Economic Conversion and Disarmament.

While it's obviously true that pieces of technology, like ball and roller bearings, are going to be used not only in military technologies but in civilian stuff as well, the fact is that major new technologies were not foreseeable as having the property of being useful for both military and civil society. As a war economy deindustrializes, part of the work goes into more military stuff, but the major part of the deindustrialization is simply the shutdown of civilian work in this country and its transfer elsewhere, mainly to countries that pay low wages and, very importantly, discourage the formation and operation of trade unions. The militarization of the economy then has two sides: the continuation and the expansion of the militarization in the U.S. and the cessation of all manner of civilian work and its transfer of the investments for this work, especially to China.

What does it mean to be in a permanent war economy?
The economy that served the military was large, diverse, elaborately equipped, well financed, and was being made a continuing part of the American economy, and a continuing interest of the federal government. The military economy has become a very durable part of the economy of the United States. But I wonder these days if it ever can be truly permanent. In a book I'm writing now (*Wars Unlimited*), I give special attention to the idea of the previously understood durability of the war economy. I don't think it's going to be that durable. I think it's going to be forced to give way.

Why was it conventional wisdom at one time that military spending was good for the economy?
This idea has its origin in the middle of the first Roosevelt administration. The history was approximately as follows. Roosevelt entered the White House in 1933. That was clearly at the bottom of the Great Depression. There were about ten million unemployed. There were bread lines in every major city. The Roosevelt administration tried a series of civilian public works, some of them emergency measures. The emergency measures started with food distribution, but it extended on to devices like the Civilian Conservation Corps, which invited young men to show up for an enlistment, so to speak, and were put to work on environmental conservation jobs that were organized in detail by the Army. But these previously unemployed young men were suddenly full-time occupied and the military made available decent food and clothing. So the work in combination with the food and clothing and medical care that became available was a big boost in the level of living for hundreds of thousands of young men who were enlisted.

As you reach the end of the 1930s, starting with late 1937 and then into 1938, there was a heating up of the military political temperatures, notably in Europe. One of the earliest things that was undertaken were programs to expand the U.S. Navy. As this was done, the economists in the United States seemed to make a discovery. The discovery was that the U.S. economy could make considerable expenditures – certainly enlarging the expenditures for the military – and it didn't detract from anything else.

Is this the guns and butter idea you talk about?
Yes. It later became known as the guns and butter theory. But it reached a new kind of peak

during the Second World War, when there was a period of massive building of new industrial facilities for the military. So there was a four or five-year stint of very concentrated work for the military, which did not take away from anything civilian. The number one reason for this was that the military work had been going on with great intensity for a relatively short period, hence the understanding of the motto that the U.S. could have both the guns and the butter.

What could the U.S. administration learn from the collapse of the Soviet economy, in terms of its long concentration of production resources on the military product?
The Soviet economy clearly collapsed by 1990 and there were the fearsome conditions of short-

The military economy has become a very durable part of the economy of the United States. But I wonder these days if it ever can be truly permanent.

age of every kind of consumer goods. During the Russian financial crisis in September/October 1998, the relatively well-off residents of Moscow felt the sting of the fall of the ruble. What was the fall of the ruble? Formally, there were six rubles equivalent to one dollar. But then the value of the ruble to the dollar collapsed and suddenly the ruble was 21.5 rubles to the dollar. And suddenly, they were confronted by the degree to which Russia had become dependent upon imports for the most ordinary goods. So that the opportunity for reconstruction after the Second World War didn't quite take hold in the whole realm of civilian goods.

How can the U.S. make good use of alternative-use committees and redesign defense-related factories, bases, and labs, to avoid something like this occurring?
During the period of lively functioning of the commission on economic conversion, we very carefully drew up a succession of plans, which finally were embedded in proposed laws. The key piece of these laws was the requirement that every military-serving enterprise was required to set up an alternative-use commission. What would the commission have as its task? First of

all, it would have to examine what alternative products the facilities and the skills of the people working in that place could be used for. They were also required to examine the market possibility for those products and what would be required for the retraining of people, if that were necessary in order to produce alternative products. The key piece was to set in motion a detailed planning process with ample time to do the work. In other words, this was not to be a crash job. This was to be done in a serious, considered way.

Who's opposing the conversion today?
In the first place, it's not being proposed by anyone in either the House of Representatives or the Senate. But I don't see this as a permanent condition. I don't see the prospect of a permanent war economy going on indefinitely. I think the damage that is now in process to the American economy is very considerable, and is going to be more visible all the time. I see a prospect, though I can't put a timetable on it, for the idea of economic conversion and thereby not only the occupational transformation but also, very importantly, the industrial economic transformation. This is to say, the United States will be moved away from a war economy.

Go civilian or go broke, right, Seymour?
Hear, hear.

Is it possible to conduct international life without a war system?
I think we'll have a better estimate of that as we see, for example, the countries of Western Europe not rushing into concentrated military production. The quality of life is high. And remember, taken together, they equal in population, and in land mass, and in value product the whole of the United States. As the conduct of life in Western Europe, with its present focus on the civilian economy, gets to be better understood on this side of the big water, it will allow for raising questions that now seem to be unreasonable or unnecessary.

Seymour Melman is professor emeritus of industrial engineering at Columbia University in New York.

Military   or   Civilian?

Cost of creating U.S. Nuclear Weapon Overkill Capacity, 1940-96	**$5.3** trillion	More than twice the net value of the plant and equipment in America's manufacturing industries
F-22 Raptor Advanced Fighter program ($228 million per plane)	**$99** billion	3,500 miles of Maglev train lines, running at 266 miles per hour
Navy SSN 774 Virginia Class Submarine program ($71 billion) + Navy Advanced Amphibious Assault Vehicle program ($8.7 billion)	**$80** billion	Investment needed to provide 20% of U.S. electricity supply from renewable and clean sources
Army Comanche Helicopter program ($48.1 billion) and Navy Joint Standoff Weapon program ($11.2 billion)	**$59** billion	Cost of building housing for the 600,000 homeless families in the U.S.
Total cost of the Navy's "Future Surface Combatant" program	**$11** billion	Annual shortfall to meet federal safe drinking water standards and replace aging facilities
Amphibious Assault Ship program	**$11** billion	Research program to develop zero emissions, coal gasification power plants
Two Navy CVN6-B Aircraft Carriers	**$10** billion	Annual cost to provide sanitary water to 2.4 billion people worldwide
E-8C Joint Surveillance Target Attack Radar System program	**$9.1** billion	Five years of funding for a global tuberculosis control program
One Global Hawk Unmanned Drone	**$210** million	Electrification of 50 miles of mainline railroad
One "upgraded" Abrams Tank	**$7.9** million	Annual cost to enroll 1,100 children in Head Start preschool programs

8.05 FROM SWORDS TO PLOWSHARES. Since the late 1960s, Seymour Melman, professor emeritus at Columbia University, has championed the conversion project – from a military to civilian economy – and has deliberated on the public good we could achieve for the money we spend on the military.

Make peace, not war: With the world now spending one million million dollars on the military per year and 54% of peace agreements breaking down within five years of signature, can we say with sincerity that we're committed to peace?

Either war is obsolete or men are.

– Buckminster Fuller, inventor, architect and philosopher (1895–1983)

In 1957, Nobel laureate Lester B. Pearson said, "The time has come for us to make a move not only from strength, but from wisdom and from confidence in ourselves; to concentrate on the possibilities of agreement, rather than on the disagreements and failures, the evils and wrongs, of the past." Nearly fifty years later, have we made that move?

The nuclear warheads are still with us, and proliferating, and the capacity of science and technology has grown exponentially since the Cold War era, lending the same tools that have contributed to national defense and the perpetuation of war to the infrastructure that now connects us all as citizens of a global network. Can we demilitarize these technologies, turn them on their heads, and prepare for peace as exuberantly as we've prepared for war?

The Information Technology, War, and Peace Project at the Watson Institute at Brown University has developed a concept called "infopeace," a corollary to "infowar," or information war. Using Gregory Bateson's definition of information as "any difference that makes a difference," infopeace seeks to make a difference about the quality of thinking: to be pragmatic, by countering a national state of war with mindful states of peace; to prevent, mediate, and resolve states of war; and to actualize peace through a creative application of information and technology. The team at the Watson Institute's Global Security program, directed by political scientist, author, and professor James Der Derian, makes critical documentaries (*After 911*, for example) and regularly plays host to conferences that encourage discussion, negotiation, and problem-solving among military and nonmilitary thinkers and organizations who support frank, serious, and complete exchanges of views.

Consider this a bold step to imagining a better world. We've made a move.

James Der Derian
on imagining peace

You talk about the military's use of information as a force multiplier. How is that?
It's a term that originally signified a sort of propaganda – psychological or psy-op – that you would have as an adjunct to the soldier in the field, officers who would provide leafleting, or even bullhorns. Consider the example we saw in *Apocalypse Now,* preceding battle by playing Wagner on your loudspeakers when the air cav is coming in. These are all forms of intimidation. Contemporary tactics have moved beyond that. It's no longer about simply increasing the effect of command and control of the battlefield. It's also about bringing to bear computers, new communication technologies, new intelligence, and multiple media, in a battle for reality in which you're shaping – in addition to the outcome on the battlefield – the opinions, beliefs, and decisions that are part of any temporary struggle that lasts longer than the usual one- or two-week international conflict. The military use of force multiplying effects is about the ever-increasing coupling of science systems and weapons systems.

Is Sun Tzu's notion of military force based upon deception now more true than ever before?
I'm sure he could only be envious of the tools we have at hand now, compared to the gongs and drums that he would use to multiply the force of conventional arms back in 500 B.C. But if you go to military doctrine now, they call for something called "full-spectrum dominance," which means using every single available technological information tool to deceive. That's deception on a tactical level. We need to also consider the levels beyond tactics and strategics, and the extent to which we have new levels of dissimulation taking place, on the levels of decision-making, how we read the images, and how the public is informed about foreign policy.

How does Paul Virilio encourage us to think about military technologies?
He gets into the empirical detail of how these new technologies emerge in a coterminous, coeval way – the way that, for instance, the machine gun and cinema are interdependent on the same technology. Or even how viewing reality for the first time out of a steam locomotive train alters our way of seeing the world. Virilio's very good about looking at how the war machine has led, unfortunately, to much of the innovation in how we see the world. In some cases, this interdependence is symbiotic. In other cases, the military is the avant-garde, the leading force. The military has taken these relatively crude technologies and refined them for the purposes of killing people. Then, in the same way in which you'd use them to prepare and execute for war, you use them to represent the war back to your populace – through the first-time use of aerial reconnaissance and high-resolution images, all the way back to the Civil War, when bodies were posed on the battlefield to make it look more real, to how people are reading the images that are now coming out of Baghdad.

Explain this notion of "virtuous war."
I like to believe it's a felicitous oxymoron, in the sense that you have this tension between people who believe you can use war to achieve ethical aims – that's the virtue part of it – and the virtual, how you can fight wars now from a remote distance and have minimal casualties on your own side. But the harm, I think – and the reason why I attempted to capture this contradiction of virtuous war – is the belief that you can use military violence to resolve intractable political problems. If you have the technological superiority, and you believe in your ethical superiority, these factors combine to a very nasty effect, which is that you defer civilian diplomatic action and give the military the opportunity to step into this vacuum and offer up solutions.

It's military policy to prepare for the worst-case scenario. They have all these incredibly meticulous computer simulations and war games and training exercises that can be taken off the shelf, while diplomats or our political leaders are wringing their hands wondering what to do.

With respect to the virtual reality training simulations for soldiers, what is the meaning of the video game once the virtual invades reality?

When I used to go out on these war games and interview soldiers at all levels of command, I'd come across some private or specialist up on a hilltop in the Mojave Desert, and I would ask, "What are you getting out of this war game?" He'd look around and make sure there was no superior within earshot and say, "I'm getting a good suntan, that's about it." There's a sense that these games aren't like reality among people going through the paces. When I interviewed the survivors of the Mogadishu raid, their line was, "This wasn't like what we trained for. We thought

> If you have the technological superiority, and you believe in your ethical superiority, these factors combine to a very nasty effect, which is that you defer civilian diplomatic action and give the military the opportunity to step into this vacuum and offer up solutions.

we'd come in, shoot them up, take casualties, and get out." It's a famous line of various generals that if we really understood how horrific war was, or is, we wouldn't go to war.

Should we be wary then about the U.S. Army and Hollywood relationship?

It's not just the Army and Hollywood. It's the Army, Hollywood, the Academy [of Television Arts & Sciences], and Silicon Valley. It's what I call the military-industrial-media-entertainment network. We know that military strategy can wag the dog of civilian policymaking. We see the same thing with the creation of incredibly high fidelity virtual realities, where the real thing starts to pale in comparison. If you have these virtual environments based on worst-case scenarios, which indicate how we're going to represent the enemy, their threats to us, and how we respond quickly, because speed is of the essence, then all of the human attributes – deliberation, empathy, and

experience – become secondary to a machinelike response. Yes, there's cause to be wary.

How do we break out of these military worst-case scenarios that hold us in a cycle of killing and destruction?

For me, this is the central dilemma: Reproducing a reality through technical means that don't allow for imagining alternatives to take hold stultifies the imagination. People say 9/11 was a failure of intelligence. Well, it was a failure of many things, including our ability to imagine beyond the confines of computer simulation.

Tell me about the idea of infopeace rather than infowar.

One of the biggest challenges facing us at the Watson Institute is trying to make infopeace a robust concept. Let's face it, war is more ubiquitous; it's sexier; it's more thrilling. This is why there are thousands of books written about war and only a handful written about peace. Trying to vitalize notions of peace means getting more directly to individuals who embody it. Why do we keep going back to Gandhi and Martin Luther King? Partly because we don't have powerful peace movements today.

There have been episodic peace movements that have been effective. The antinuclear movement had a profound effect. The antiwar movement for Vietnam did also. We're still working on the anti-interventionist, anti-terror movement. So there will be an increasing interest in having a critical tool. Infopeace is a tool we're trying to develop through the multiple media of video documentaries and video teleconferences. We make efforts to bring together people at the Watson Institute to get into a dialogue – military officers, the media, academics, and nongovernmental organization (NGO) activists. Finding a common language is usually the first step, but you'd be surprised at how, at the end of two or three days, people are intensively interacting. Real shifts occur. Not radical worldview transformations. But the dialogue has certainly begun.

James Der Derian is the director of the Global Security Program at the Watson Institute for International Studies at Brown University in Providence, Rhode Island.

We will eliminate the need for raw material and banish all waste.

Edward Burtynsky. *Oxford Tire Pile #1, Hamilton, Ontario, 1997.*

The idea of the endless cycle of design and production promises a shift in manufacturing processes from the wasteful industrial systems of the nineteenth and twentieth centuries. The new design model provides a continuous assembly/disassembly line that cycles the product and its constituent matter – rather than recycling it – in a never-ending loop of improvement.

MANUFACTURING ECONOMIES

Edward Burtynsky. *Urban Mines, Densified Oil Drums #4, Hamilton, Ontario, 1997.*

Edward Burtynsky. *Urban Mines, Phones #21, Hamilton, Ontario*, 1997.

BARBIE BLVD

9.01

9.02

FROM TAKE-MAKE-WASTE TO NO WASTE. Barbie paraphernalia and pink plush toys relentlessly accumulate (9.01) while the Ford River Rouge facility supports a new cycle-to-cycle protocol, where waste is continually fed into the system as a food source. Housed on its roof (9.02) is the world's largest habitat, according to *The Guinness Book of World Records*.

9.03

9.03 BIO-ALCHEMY. Eco-designer John Todd uses natural cycles to the utmost. He cleans up water systems around the world and can transform brewery waste into gourmet mushrooms and greens. He says, "The waste becomes fish feed — and then, throw in some earthworms — and you have rich soil to grow mixed greens. Once the greens are harvested, you're left with a powerful organic fertilizer to grow mushrooms." And on you go. The system even collects heavy metals out of the waste stream.

Cycle to cycle: Instead of disposing of waste, think about how to use it as an input. The goal is no waste generation at all. Apply the intelligence of nature to human needs. Waste = food.

Each thing is of like form from everlasting and comes round again in its cycle.
– Philosopher-emperor Marcus Aurelius (121–180 A.C.E.), *Meditations, II, 14*, 167 A.C.E.

Major manufacturers are leading the way out of the old into what architect William McDonough and partners call "The Next Industrial Revolution." Nike's supply-chain-integration tools enable it to monitor all the material throughout its entire supply chain. The Herman Miller furniture company has just released its first cradle-to-cradle chair – no PVC – that could become a chair again and again ad infinitum. In 1993, Steelcase Corporation created its first fabric with McDonough Braungart that's clean enough to eat. The Ford Motor Company is developing the Model U (which follows in line after the Model T), a cradle-to-cradle car whose materials can go back to the soil, or back to industry, forever. BASF Chemicals, the largest chemical company in the world, is developing polymer protocols for infinite recycling of intelligent chemistry. And Instituto de Empresa in Spain – the most impressive business school in Europe, according to *The Wall Street Journal* – has inaugurated a new program for eco-intelligent cradle-to-cradle management.

Vermont-based ecological designer John Todd creates what he calls eco-machines. He holds a Ph.D. in Fisheries/Oceanography and is trained in agriculture, parasitology, and tropical medicine. His integrated networks of microorganisms, higher plants, snails, and fish systematically process waste through an ordered and natural assembly/disassembly line, removing specific toxins from water along the way and leaving only a vibrant ecosystem behind. He and his wife, Nancy Jack Todd, founded Ocean Arks International in 1981. Their motto is "To Restore the Lands, Protect the Seas, and Inform the Earth's Stewards..."

9.04

9.05

9.06 PVC FREE. Herman Miller's new Mirra design is a high-performing, environmentally advanced work chair that's easy on the back – and the environment.

9.04 and **9.05** DESTINED FOR THE DISASSEMBLY LINE.
The Model U is a concept car by Ford that promotes materials – such as eco-effective polyester – that are safe to produce, use, and recycle over and over again in a cradle-to-cradle cycle. Ford asserts that these materials never become waste, but instead are nutrients that either feed healthy soil or the manufacturing processes without moving down the value chain.

William McDonough on economy, ecology, and equity

You talk about The Next Industrial Revolution, where industry and environment come together in harmony. What does this look like?
It looks at the idea, as Francis Crick said in 1962, that in order for something to be vital it has to have growth, it has to have a free form of energy, and it has to have an open system of chemicals. So if we think about a tree, it has to have some cells that grow, even for simple reproduction, and it has to have free energy from outside the system, in this case natural sunlight, and it needs an open system of chemicals that synthesize within its metabolism for the benefit of the organism, its reproduction, and its ecosystem.

If we saw human industry in a similar way, we'd realize that there's something relatively new in evolutionary terms that we call technical nutrition. Not just biological nutrition, which is the living thing powered by the sun and consumed by other organisms as they breakdown (or, as we say, "waste equals food"), but actually seeing human artifice and technology as something that is put into the same kind of cycle. These are what we call technical nutrients.

Take aluminum for example. Our species has made 680 million tons of aluminum since 1880 and we still know where 440 million tons are. So the idea would be that you would design two kinds of things: products of consumption, those things that are literally biologically consumed and go back to soil, or products of service, things from which we want the service, but not necessarily the molecular potential. With something like a computer or a car or carpet, the user is a customer not a consumer. These are services and in fact, when you finish with a synthetic carpet, for example, you should be able to either return it back to industry forever and remake carpets or other useful things. So biological and technical nutrition – that's the protocol we initated and have been continuously championing and developing.

What is the difference between eco-efficiency and what you call eco-effectiveness?
Eco-efficiency (doing more with less) as a strategy is well meaning but not necessarily adequate to the task. Being efficient means that you're probably doing something right, in terms of using the least to do the most, but the problem is that if you're doing the wrong thing, it might be pernicious because it perpetuates the wrong system with the erroneous thought that things are getting better. For someone to tell a company to be more eco-efficient and to please make twice as many cardboard boxes out of the trees in Indonesia, it sounds like a factor 2 efficiency. Even if they said make it factor 4 or factor 10, you still haven't really solved the problem, because it's still good bye to Indonesian forests. Why would you use something as beautiful and as diverse as a tree for something as prosaic as a cardboard box that's used once or even twice and then put into a chlorine-laden recycling loop that is actually continuously down-cycling all the materials and destroying water quality?

From our design perspective, the question really needs to be, "With eco-efficiency, is being less bad being good, or is it simply being bad, just less so?" With eco-effectiveness, on the other hand, we ask the question, "Am I doing the right thing?" And then we start to do it efficiently, so we can create prosperity and growth.

So we're not interested in being less bad. We're interested in being 100% good?
Right. That means you have to design with positive principles and positive goals. Modern industrial culture doesn't seem to have principles, except something like: "If brute force isn't working, you are not using enough of it." While its goals are unclear, its de facto goal appears to be to create ecological and human tragedy. If you play a game, you have to have a clear goal; in chess, you're going to take a king. So we have an endgame in mind because without this, strategy becomes meaningless. What we seek is a delightfully diverse, safe, healthy, and just world, with clean water, air, soil, and power, that is economically, equitably, ecologically, and elegantly enjoyed.

How did you get turned on to the idea of changing the world of design?

I grew up in Hong Kong, so I was in a place with four hours of water every fourth day during the dry season and six million people on forty square miles. I saw a lot of optimization of very precious resources. Then, as we went to the Pacific Northwest for the summers with my grandparents I saw astonishing abundance: fresh water, big forests, pure springs, salmon. I went from a world of extreme limits to

Modern industrial culture doesn't seem to have principles, except something like: "If brute force isn't working, you are not using enough of it." While its goals are unclear, its de facto goal appears to be to create ecological and human tragedy.

a world of extreme abundance, and yet my grandparents were also very careful and kept the spring clean, composted organic waste, and saved rubber bands and aluminum foil and so on. I always thought the world was something you took care of, and it hopefully got better because you were there. I also saw in Chinese agriculture a perpetual agriculture: farmers for forty centuries farming the same piece of ground. So that was the context in which I grew up. When I came to the United States to live as a teenager, I entered a world of profligacy and seeming wanton abandon of things in a take-make-waste production system, with a cradle-to-grave "throw it away" philosophy. I think this, in many ways, was the result of nuclear threat, something I did not live through. While I was a child in Hong Kong, third-graders in the U.S. were being taught how to dive under their desks because Armageddon may appear at any instant. When you sense that everything

could end in an instant, you live as if there might not be a tomorrow. This became embedded in the culture – modern culture actually created geopolitical and physical threats (global terrorism, weapons of mass destruction, biological warfare) that could destroy us all tomorrow – so many industrialized countries have a "get it while we can" attitude rather than a continuous longterm prosperity in mind.

What goes on in a cradle-to-cradle cycle?

Cradle-to-cradle essentially says that you have an open metabolism of chemicals that are manifesting benefit for living systems or technical systems. They're not contaminating each other and they are designed to either replace themselves in cycles or get better as they go through the system. Typically what we call recycling today is down-cycling in our lexicon. Things are actually getting lower in quality as they go through the process. Clear milk jugs will be transformed into a park bench that's on its way to a landfill or an incinerator, getting contaminated by various additives and dyes and losing its quality through the system. We've been looking at nylon fibers, for example, that can be chemically recycled, and actually up-cycled. They get better as they come back and go through the new cycle because mechanical properties have been improved, thereby increasing the quality of the fiber. Essentially, cradle-to-cradle says that if things relate and can improve soil health, then we may return them to soil.

What does your triangle diagram mean to you?

We use this triangle known as the Sierpinski gasket, or fractal tile, to be able to navigate the relationships between ecology, equity, and economy. It's a fractal way of looking at the entire universe that's self-similar. Cost, performance, and aesthetics meet life, liberty, and the pursuit of happiness!

William McDonough is an architect and co-author with Michael Braungart of Cradle to Cradle: Remaking the Way We Make Things.

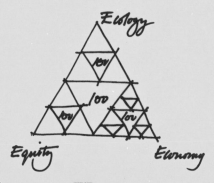

9.07

9.06 FITS LIKE A GLOVE. ILC Dover is a world-leading firm that specializes in the engineering of soft-goods products. They use stereolithography as the pattern for final tooling in the rapid manufacturing of space gloves for NASA astronauts.

RAPID PRECISION. As the pattern for final tooling, stereolithography is faster, cheaper, and more accurate than conventional manufacturing methods. Seen here: a rapid-prototype pump (9.07) and engine block system (9.08).

Atoms to bits to atoms: Powerful new tools for virtual simulation link the entire manufacturing process in a continuous flow – from design to prototyping, to fabrication, and even to disassembly. The capacity in one field is immediately applied to another.

Integrating computer-aided design with computer-aided fabrication and construction fundamentally redefines the relationship between designing and producing.
– Malcolm McCullough and William J. Mitchell, *Digital Design Media*

As we transition out of the Industrial Revolution into the Next Industrial Revolution, we move not only from a take-make-waste to waste-equals-food manufacturing methodology, but also from atoms to bits and back to atoms.

In 1981, Dassault Systèmes of France developed a software product known as CATIA. It allows manufacturers to simulate all the industrial design processes – from the pre-project phase through detailed design, analysis, simulation, assembly, and maintenance – and has since become the most commonly used product development system in the world. It is the benchmark tool in the automotive, aerospace, shipbuilding, plant design, electrical and electronics, consumer goods, and fabrication and assembly sectors. Dassault Systèmes, Boeing, and IBM worked together in the design of the Next-Generation Boeing 737. The entire process of designing, building, and even aerodynamic testing of an airplane was reconceived. Data management, user productivity, and visualization were enhanced – all three of which were necessary to effectively manage the size and scale of millions of airplane parts on a computer screen. CATIA gave Boeing engineers and designers the ability to see each part as a solid image, then simulate the assembly on-screen, easily correcting misalignments and other fit or interference problems.

CATIA has also opened up what's possible in architecture. Frank O. Gehry uses the computer-modeling capabilities of CATIA in the development of his structurally complex buildings. Without it, the construction of the Experience Music Project or Walt Disney Concert Hall would not have been possible.

9.09 ARCHITECTURE WITH A TWIST. In 2002, Frank O. Gehry developed Gehry Technologies LLC, a Gehry Partners spinoff company. Its mission is to take the CATIA platform to the wider architecture, engineering, and construction (AEC) community by leveraging the very best in software tools and processes for architectural design and building construction projects. This is a move that promises to revolutionize how buildings, such as the Walt Disney Concert Hall (below), get made.

9.10 SIMULACRA AND SIMULATION.
CATIA revolutionized the entire process of designing
and building the Next-Generation Boeing 737 airplane.
It gave Boeing engineers and designers the ability
to see each part as a solid image, then simulate the
assembly on screen, easily correcting misalignments
and other fit or interference problems.

We will design evolution.

A needle containing human DNA in solution is injected into a fertilized mouse egg.

When Franklin, Crick, and Watson discovered the structure of DNA in 1953, the realm of the living was rendered as a system of information. Since then, we've grown in our capacity to explore every aspect of life as we know it – from biological systems and products to new forms of intervention in medicine and genetic engineering. As Alvin Toffler wrote in *The Third Wave*, "Second Wave thinkers conceived of the human species as the culmination of a long evolutionary process; Third Wave thinkers must now face the fact that we are about to become the designers of evolution."

Matt Ridley
on the genome

How significant is it that for the first time in four billion years a species on this planet has read its own recipe?
We don't entirely know how to understand the significance of this, but we have just in the last year, for the first time, got an absolutely nailed-down, gold-standard sequence of what's in that. It's a very, very big document – as long as eight hundred copies of the Bible [inside the nucleus of every cell] – and it's written in DNA code, which consists of a four-letter alphabet. It's linear, digital, and just like text. We know in a sense that even with a 26-letter alphabet we could never exhaust the number of potential books that could be written, and that's what genomes are all about.

What significant scientific work was going on just before the discovery of the double helix?
It was a wonderful period, fifty years ago, with the birth of molecular biology. And in retrospect it all fits together. There were a whole series of steps that led to the discovery of what the gene was made of. Everybody knew what genes were; everybody knew that inheritance came in particles in some sense. There were blobs of inheritance: you either got blue eyes or you got brown eyes; you didn't get something in between. That was what Mendel discovered. By 1944 anyone who was in the know knew that genes were made of DNA. That was because of a series of brilliant experiments by a man named Oswald Avery, who never gets quite enough credit, who pinned down

that DNA was the substance of which all genes are made. But nobody could figure out how, because DNA seems to be a monotonous and simplistic chemical compared with proteins, which have a lot more diversity.

If you go back to 1953 and ask who was predicting how DNA would have this capacity for carrying inheritance from one generation to the next, they were all barking up the wrong tree. They were talking about something to do with special quantum energy states; they were talking about special three-dimensional configurations. In fact, it turned out that it's a simple linear digital code. In other words, there are four chemicals and they're repeated in a significant order, which gives you a piece of text that tells you whether your eyes are blue or brown.

On the 28th of February, 1953, at 9:30 in the morning, it became immediately obvious that what we were talking about was a digital sequence. This is the time when Jim Watson found the base-pairing phenomenon. He discovered that these letters fitted together on the opposite strands of DNA in such a way that A and T fitted together with the same shape as C and G. It really was a "Eureka" moment.

Up until recently, we didn't hear much about Rosalind Franklin. What role did she play?
A very important one. She arrived on the scene in late 1951, taking over the project from Maurice Wilkins, who had developed a technique for taking X-ray photographs of DNA. Franklin perfected this technique and managed to get a photograph that showed what shape the molecule was. Wilkins had suspected that it was helical, and she proved it beyond doubt. At this point, Watson and Crick started playing with models, and they solved the problem. Franklin could have done it though, as she had the best data. If she had done it all alone, she would have gained it for Britain, for women, and for Jews. It would have been such a great story. So, in a way, I feel frustrated with her rather than sorry for her – that she didn't manage to grasp that prize. But there were lots of reasons for that, including institutional sexism, which has clouded the history of the discovery of DNA.

How are old ideas and models changing as a result of the new genomics?
This is a science that revises itself continuously. A lot of early guesses about genomes have proved to be wide of the mark in interesting ways. That's not to say that things are proving the opposite of what was proved before. But, for example, an

early guess was that there would be about 100,000 human genes. That's 100,000 discrete paragraphs of text in the genome that actually spell out the recipes for protein. And that seemed to be a pretty good guess, which was based partly on the idea that the human brain is so complex, it would require a whole bunch of special genes that weren't found in other animals. It turns out that this is wrong, in a big way. We have 25,000 genes, not 100,000. That's 6,000 genes more than a microscopic worm has, and 15,000 less than a rice plant has. Far from being the most sophisticated machine on the planet, the human being is just an ordinary creature. Our species is just another twig on the evolutionary bush.

On the 28th of February, 1953, at 9:30 in the morning, it became immediately obvious that what we were talking about was a digital sequence.

What are your thoughts on the hotly debated golden rice?
The man behind golden rice is Ingo Potrykus, a Swiss biotechnologist who had the desire to apply genetically modified plant technologies to solving some real Third World problems. In his research, he found that the most solvable problem was with vitamin A deficiency in rice. Something like 500,000 children a year suffer from sight problems as a result of vitamin A deficiency. So he asked, "What if we could just take a gene out of a daffodil plant and insert it into rice?" And he did it, on his own, not under the aegis of any corporation. He managed to persuade all the companies that had the patents that he'd breached to give him a license for free. He can grow this stuff in his laboratory, put it in his pocket, fly to Jakarta, and hand it out if he wants. There's nothing to stop him in the law, except for the fact that all the countries where he'd like to distribute it have passed laws, under pressure from organizations like Greenpeace, who disapprove of this technology, saying that he shouldn't be allowed to do this because the risks of it being a genetically modified kind of rice have not been properly assessed. I think it's pretty clear where right lies in that argument and where wrong lies. And, you know, he's been vilified by a lot of the environmentalists, and quite wrongly so.

Who knows, maybe one day it will be an example of the actual future proving to be far better than the forecasted hype of a future?
This is one of my great themes and holds true if you go back and look at the number of times people have predicted disaster, from Malthus at the end of the eighteenth century onwards. A very nice example was in a speech to the British Association for the Advancement of Science in 1898, when the scientist Sir William Crookes said, "England and all civilized nations stand in deadly peril of not having enough to eat." He predicted that we were all going to starve. Along came the Haber-Bosch process [a method of directly synthesizing ammonia from hydrogen and nitrogen, developed by German physical chemist Fritz Haber], which is how we make nitrogen fertilizer out of the air. The organic movement was a reaction against the use of inorganic nitrogen fertilizer. But if we hadn't done that, then we couldn't possibly feed six billion people, which is what we're managing to do now, and leave room for rain forests. The productivity of land without artificial nitrogen just isn't there. Again and again, technology comes to the rescue of these environmental problems rather than causing them.

What was it that James Watson said – that the problem is not biotechnology itself but the slow pace in which we're applying it to solve some of our world problems?
Yes. Slamming on the brakes on a technology is not a morally free route to take. There are moral hazards in taking that course because you might be preventing a new technology from coming to the rescue of people with real problems. The sooner we can get some of these molecular technologies out there, the sooner we can get the cure for cancer organized through molecular biology, which is where it's going to come from, and where it's already coming from in some cancers. Then the better the future generations are going to be. In that sense, I'm a sort of techno-optimist and I'm sometimes called a bit of a Dr. Pangloss.

Matt Ridley is an award-winning author based in Newcastle-upon-Tyne, England.

10.01 GENE MARKERS. CAMBIA's scientists use Green Fluorescent Protein (GFP), a nontoxic marker used to label cells expressing transgenes, and their own beta-glucuronidase (GUS) reporter system to label novel traits that exist in the genetic pool of transgenic rice.

Designer genes: Genetic engineering has great power and could have the potential to narrow the gap between the developed and developing worlds. It is up to the global community to decide how it is ultimately used.

What are the consequences of reducing the world's gene pool to patented intellectual property controlled exclusively by a handful of life-science corporations?

– Jeremy Rifkin, author of *The Biotech Century*

Biotechnology, in its broadest sense, has been around since humanity began harvesting crops, breeding animals, and brewing hops – all examples of our adventure with natural phenomena and our application of lessons learned from them, for human benefit. However, the most astounding developments came out of the molecular biology revolution that was kick-started with the discovery of the double helix in 1953.

Three key moments between then and now include, first, the creation of recombinant DNA organisms (by Stanley Cohen and Herbert Boyer); second, the formation of Genentech, the first biotech corporation (co-founded by Boyer), specializing in commercial products using recombinant DNA technology (Genentech cloned the insulin gene in 1978); and third, the U.S. Supreme Court ruling in 1980 that live, human-made organisms are patentable material. Between 1980 and the present, bold pronouncements on human cloning emerged (the Raelians and Richard Seed); fantastical experiments followed (Steen Willadsen's "chimeras" and Ian Wilmut's Dolly); and big business clamored to claim ownership over millions of acres of transgenic crops.

Richard Jefferson, founder of CAMBIA (the Center for the Application of Molecular Biology to International Agriculture), is launching the open-source-modeled BIOS initiative to decentralize and democratize agricultural biotechnology. Through public debate, creative shared technology development, and reformed public policy, his mission is to enable sustainable agricultural innovation. This includes open exploration into the repertoire of plant genomes and the freedom to harness the latent diversity within them.

Freeman Dyson
on genetic engineering

Freeman, what are some of the more outstanding scientific breakthroughs you've personally lived through?

The most outstanding, of course, was the double helix, the discovery of the structure of DNA. As soon as we saw that little two-page article in *Nature* in 1953, I think we all recognized that this was the big step forward. And it has been, I think, the big event of the last fifty years.

Your inventions for the benefit of all humankind involve freeze-dried fish, warm-blooded plants, and even turtles with diamond-tipped teeth. However, I'd like to hear more about the silicon leaves.

Yes, these all relate to genetic engineering. This is a hugely powerful technology but it's not something that just comes suddenly into the world. It has to be developed slowly and carefully over long periods of time. The point about genetically engineered leaves is simple. Let me explain. Today we have two ways to use solar energy. One is to manufacture silicon collectors that turn sunlight into electricity with ten percent efficiency. The other is to grow plants such as sugarcane that turn sunlight into chemical fuel, with one percent efficiency. The first method is too expensive to compete with fossil fuels. The second method uses too much land. The question is whether by using genetic engineering we couldn't design crop plants to use sunlight with ten percent efficiency. In other words, could we grow trees with silicon leaves instead of green ones? I don't know why not. This way, the leaves would become solar collectors and we would have a cheap way to use solar energy using only a tenth as much land as existing crop plants. If it could be done it would transform the world and spread the wealth much more evenly over the earth.

What does the biotech industry today share in common with the nuclear industry in terms of public perception?

I think the public is rightfully scared of both. The reason the public is scared of the nuclear industry is because it's also associated with bombs. I think the public is distrustful of the biotech industry because some of its first applications were putting poisons into food – that is, putting pesticides into crop plants. That was a tactical mistake just as it was a tactical mistake of the nuclear industry to build the bombs first. There are other things you can do with biotech, of course, like producing food with much higher nutrient value or producing plants that will grow in poor soils or in unfavorable conditions, which is what the world badly needs.

I like the idea you had with the Orion project in the 1950s: you designed a huge nuclear-powered spaceship with the view to take nuclear weapons out of the hands of the military.

It's true. The idea was to transport nuclear weapons into outer space and detonate them before the military could. Right at the beginning of the space age we had this idea we could use nuclear bombs to drive a really big spaceship all over the solar system. We never built it, but in principle, it would have worked and we could have gone sailing around Mars and Jupiter and Saturn, which is what we intended to do. I was very serious about that. It never got the green light for various reasons. It turns out, of course, that if you explode thousands of bombs, you make a great deal of radioactive contamination. It's not good for the local ecology – even in the far reaches of our solar system. That was the main reason why we didn't do it in the end.

Explain the work you've done with adaptive optics.

Adaptive optics is a new kind of telescope, which has been talked about for about forty years and now finally is getting built. You make very rapid changes in the shape of a mirror in order to compensate for the rapid changes in the atmosphere. If you have an adaptive optic system on a telescope, then you can actually correct for the distortions produced by the atmosphere at the rate of about 1,000 times per second; it has to be done very fast to keep up with the rapid motions of the atmosphere. This way, a telescope on the ground could have as good a resolution as a telescope in space. Instead of having one Hubble telescope in space, we could have a lot of ground-based telescopes, much cheaper and with equal resolution. The French, in fact, are leading the world in this game. They have put the adaptive optics system on the European telescopes in Chile and these are working quite well.

The question is whether by using genetic engineering we couldn't design crop plants to use sunlight with ten percent efficiency. In other words, could we grow trees with silicon leaves instead of green ones? I don't know why not. This way, the leaves would become solar collectors and we would have a cheap way to use solar energy using only a tenth as much land as existing crop plants.

The Dysonsphere, which of course made you so famous among "Trekkies," is a wonderful concept. How did you dream this up?
My idea was to look for intelligent aliens. Many people have been looking for intelligent aliens in the sky for a long time; we haven't found any yet. Normally, the way to look for intelligent civilizations way out beyond the solar system is to look for radio signals. If the aliens are intelligent they quite likely communicate by radio and they might even be wanting to communicate with us, so they might be sending radio signals that we can listen to. Forty years ago, it occurred to me that it would be interesting to look for noncommunicating aliens. Suppose the aliens don't want to communicate — which is certainly quite likely — then how can you detect them? I thought you could still detect them by looking for heat radiation, as any advanced civilization with a big technology and a large population would have to get rid of waste heat, which is radiated into space in the form of infrared radiation. So the thing to do would be to look for sources of infrared radiation in the sky.

This Dysonsphere, as it's come to be known, is not at all likely to be a big round ball. It's much more likely to be a cloud of objects orbiting around a star. I believe it's still worth looking for.

When you find yourself arriving upon a piece of elegant mathematics, do you wonder if it might be a clue to the universe, or maybe even the origin of life?
What I do is calculate particular problems. I don't think in terms of general laws or grand ideas. I look for little problems that I can solve with mathematics, and then when I find something that works, I try to carry it as far as I can. That's what I did with [Richard] Feynman's physics and also what I've been doing with the problem of the origin of life. I haven't solved the problem of the origin of life. Nobody knows how life originated. I made a little mathematical model that I can play with which may have some relevance, or it may not. Essentially, I build little models and play around with them.

With all of your knowledge and life experience, you're still imagining a better world through applied science. What of your insight on the future can you leave us with?
We still have to face the problem of the genetic engineering of humans. This is a much more delicate and dangerous subject than the genetic engineering of crop plants, but it's something we have to face. It's right and proper that the world is paying a lot of attention to it; it's going to be the main subject of our next hundred years, in my opinion.
Essentially, the question is, "Will you allow the parents to decide what kind of babies they want to have?" That's really the problem. It's not so much a scientific question; it's a question of the human rights of parents compared with the social problems of society as a whole. We have to somehow find equilibrium to give the parents some freedom, but not too much.

Freeman Dyson is a mathematical physicist at the Institute for Advanced Study in Princeton, New Jersey.

10.02 and **10.03** RICE MAPS. Susan McCouch and her team in the Department of Plant Breeding at Cornell University study ancient maps of the molecular variety and use markers to identify yield-enhancing genes among rice plants' thousands. The purpose is to better understand how certain combinations of naturally existing genes work together in a synergistic way to generate novel plant forms and functions. Called "smart breeding," this strategy promises to keep the seeds in the hands of farmers throughout the rice-growing world and generate increase yields of 3% to 5% per year for the next 20 years. At right, *Oryza sativa*, a modern cultivar of rice (10.02), and wild ancestor *Oryza rufipogon* (10.03).

10.02

10.04

10.03

10.05

10.04 and **10.05** ONE POTATO, TWO POTATO. Agricultural plant pathologist Florence Muringi Wambugu, a strong proponent of biotechnology for Africa and developing countries, hopes to alleviate hunger and poverty by establishing an operational crop biotechnology in Kenya and sharing the transgenic technology with other African countries. Wambugu, supported by Monsanto, specializes in genetically engineering crops for protection against viral disease. At left, a virus and weevil-infected sweet potato with a loss of 20% to 80% (10.04) and a "healthy" genetically engineered sweet potato (10.05).

10.06

10.06 GOLDEN OPPORTUNITY? Ingo Potrykus, of the Swiss Federal Institute of Technology, and Peter Beyer, of the University of Freiburg, Germany, developed "golden rice" to address the rampant vitamin A deficiency in the developing world. It is a food research product, in which a delivery system for vitamin A (in the form of beta-carotene) is designed to eliminate blindness in 500,000 children a year. However, according to the Rockefeller Foundation, it may deliver only a small fraction of the needed vitamin dose. An adult would have to eat nine kilograms, or twenty pounds, of cooked rice per day to get the necessary dosage.

10.07 ASEXUAL SHEEP. Embryologist Ian Wilmut and biologist Keith Campbell created "Dolly" (after Dolly Parton) in 1996, the first cloned mammal from fully differentiated adult mammary cells. Wilmut, now at Roslin Institute in Edinburgh, "the leading centre for animal biotechnology," hopes to produce animals that act as manufacturing plants for valuable human proteins.

10.08 UNNATURALLY COOL. It turns out that chickens in the tropical zones of the developing world are losing body mass on account of the heat. As a result, Israeli researcher Avigdor Cahaner cross-breeds feathered chickens with already balding birds (with the "naked neck" gene) to create featherless birds, which thrive in these same conditions. Cahaner's method, unnatural selection, is faster than natural selection but much slower than modern genetic engineering techniques, like transgenics and cloning.

Robert Langer
on tissue engineering

You are one of the most prolific inventors in medicine today. Tell me about some of the patents you hold.

We have quite a few patents in different areas. Some of them, for example, relate to polymer systems that continuously release molecules of a certain size. Another is on brand new kinds of plastics. Another is on the local delivery of drugs, for cancer or other diseases. Another is on plastics that can change their shape. Another relates to tissue engineering. Some new ones are in gene therapy delivery.

What are some of the applications of your research into the delivery of drugs at steady, controlled rates?

There are quite a few, and, of course, the field goes far beyond what we do ourselves in the lab. But some drug delivery systems you might see are transdermal patches – patches that could deliver nitroglycerine or nicotine, to prevent people from smoking cigarettes. Drug delivery also involves drug-eluting stents, where a polymer is placed on the outside of the stent; these are like Chinese finger puzzles, and they're put in a patient to keep the blood vessels open. We've been involved in implants, or injectable microspheres, that could release human growth hormone, or hormones that might be able to aid in treating prostate cancer.

When you first presented your ideas to the medical community about polymer systems that release molecules into the body, how were they received?

A lot of people felt that substances of only a certain size could move through plastic. And I needed to figure out a way to get substances of a much bigger size through the plastic. I finally figured out a way to do that, and people just didn't think it was possible. Conventional wisdom told them that trying to move through plastic was like, say, you or I trying to pass through a brick wall. See, the thing is, you could have a brick wall, or you could have something like Swiss cheese. Nothing could get through the brick wall, of course. Things will get through the Swiss cheese, certainly, but too quickly. What we were able to do was make something that had open pathways, but with tortuous routes that were narrow and incredibly winding, so it took a very long time to get through them.

How did you come to develop the chemistry microchip?

I got this idea about ten years ago from watching a television show on how computer microchips were made. I thought, rather than have an electrical microchip for the computer industry or your television, maybe we could make a chemistry microchip. I thought we could build into a little chip thousands of little wells that could have, literally, thousands of different drugs if you needed them, or thousands of different doses – like a pharmacy on a chip – that could one day give the patient whatever he or she needed at the appropriate time.

How is the drug released from the microchip?

If you make what we call an "active chip," you can control it by telemetry, the same way you open up a garage door, by radio frequency. You have a little device in your car, in the case of the garage door; you could have the same kind of device in a wristwatch or something you can hold in your pocket, which could open up the specific well in the chip containing the specific drug or the specific dose of drug that you'd want.

What thought processes do you go through when it comes to designing biomaterials?

I try to ask the question, "What do you really want in a biomaterial, from an engineering standpoint, from a chemistry standpoint, from a biology standpoint?" And then I ask, "Could you design it from scratch, from first principles?" That's what we've done. We've asked different chemistry questions to see if we can actually make these kind of things.

How does it feel to know that your polymer wafers are going into people's brains?

What's great is that I've met patients who have been treated with this innovation and it's a great feeling for everybody that has participated in the project – in the lab, and myself – to see that these wafers have provided a way to relieve suffering and prolong life.

How does the wafer work?

We designed a special polymer from scratch, which dissolves the way a bar of soap does. The

How is tissue engineering different from other fields of cell culture?

Doing cell culture alone, you can study things and learn things about cells and culture, but you wouldn't make a new tissue. Tissue engineering, if successful, ultimately enables you to make a new tissue, like new skin that you could actually put on a patient or, someday, new bone or cartilage or blood vessels or heart muscle. These are often three-dimensional structures that are vascularized. They sometimes have multiple cell types, so they are more complex structures to engineer.

We wondered if we could design an artificial liver, then came up with the idea of using a combination of polymers and cells in certain configurations. This line of thinking ultimately led to the design of new tissues.

How does a biodegradable polymer scaffold function?

Say you'd like to make a particular structure, like an ear or skin. You first take the polymer and form it into the desired shape, then you take the cells from the patient, or stem cells, and place them on the polymer fibers. These cells grow and ultimately form the structure that you want them to form. Finally, these new tissues are put into the body and the plastic scaffold degrades away.

chemistry that we've made is such that we can make it last anywhere from a day to over six years, or any time in between. In particular, what we've done is make it last for about a month. The neurosurgeon who treats the patient (we've done this a lot with neurosurgeon Henry Brem, at Johns Hopkins) operates on the brain, removes as much tumor as possible, and then inserts these little wafers, the size of small coins. These wafers slowly dissolve right at the site of whatever remaining tumor exists, and they locally deliver the chemotherapy. In contrast to conventional chemotherapy, which goes all over your body, causing a lot of toxicity to different organs, this is local delivery, right to the site of need.

Go back to the days when you and Dr. Jay Vacanti first tossed around the idea of actually growing new organs from scratch.

Jay has been a good friend of mine for a long time. About twenty years ago, when he directed the Liver Transplant Program at Boston Children's Hospital, he came to see me because he wanted to solve the problem of too many people dying while awaiting new livers. We started talking about how we could come up with a way to overcome the problem of donor shortage. We wondered if we could design an artificial liver, then came up with the idea of using a combination of polymers and cells in certain configurations. This line of thinking ultimately led to the design of new tissues.

Where can some of this work lead?

We hope that one day we might be able to take a patient with a damaged spinal cord and make him or her well again, or at least improved by these technologies. We've worked out ways to use polymer scaffolds where you could actually help patients; in this case the patients are rats, but basically we're able to take rats that couldn't walk and then a hundred days later find them actually walking.

In the next ten years, what do you have to look forward to in both drug delivery and tissue engineering?

In drug delivery there are already a lot of products helping patients, and I think we have the opportunity to see many, many more of these. You'll see more patches and new aerosols, which will take the place of insulin shots – or at least lessen the need for them. You'll see all kinds of new advances in drug delivery. With tissue engineering, we're just now at the point where new skin can be made for patients, and I think before long we'll see new cartilage for patients too, maybe even new bone and organs.

Robert Langer is a biochemical engineer at the Massachusetts Institute for Technology in Cambridge, Massachusetts.

10.09

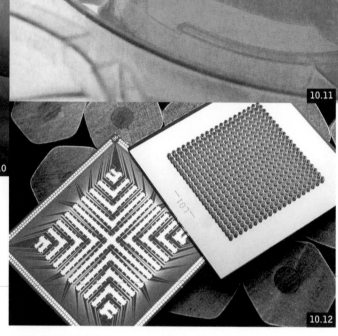

10.11

10.10

10.12

10.09 BRAIN WAFERS.
Biochemical engineer Robert Langer's chemical engineering background has also contributed to innovative work in the realm of modern drug delivery. Seen here: biodegradable dime-sized polymer wafers, designed to administer on-site chemotherapy to the brain, being inserted during tumor removal.

10.10 BIORUBBER.
Yadong Wang, assistant professor in the Department of Biomedical Engineering at the Georgia Institute of Technology and Emory University, developed a biodegradable and biocompatible elastomer while he was a research associate in the Langer lab at the Massachusetts Institute of Technology. Called "biorubber," it is strong, biocompatible (nontoxic), and inexpensive (made in large batches from commodity chemicals). Future applications may include such things as nerve guides, tissue-engineered lungs, heart valves, cardiac muscle, blood vessels, drug delivery, and medical devices.

10.11 SYNTHETIC SKIN.
Apligraf is a bioengineered skin material made by Organogenesis Inc. for use in cases of venous leg ulcers and diabetic foot ulcers. Like human skin, it consists of living cells and structural proteins. Unlike human skin, it does not contain structures such as blood vessels, hair follicles, or sweat glands. It was approved by the FDA in 1998.

10.12 PHARMACY ON A CHIP.
John Santini and his team at MicroCHIPS, Inc., a company founded out of MIT with Robert Langer and Michael Cima, developed technology based on tiny implantable silicon or polymeric microchips for the purpose of precision drug delivery. Seen here: a 15mm by 15mm silicon and metal microchip, with a layer of glass containing 400 micro-machined wells, in which any combination of drugs, reagents, or other chemicals can be stored.

Bio-polymers and biotech body parts: Although the body's own wisdom still reigns supreme, powerhouse teams of surgeons, chemical engineers, materials scientists, and genetic researchers are devising novel ways to deliver drugs and regenerate human tissues.

Ten millennia ago the development of agriculture freed humanity from a reliance on whatever sustenance nature was kind enough to provide. The development of tissue engineering should provide an analogous freedom from the limitations of the human body.

– David J. Mooney, Chemical engineer

Tissue engineering is a burgeoning field of scientific exploration that marries the principles and practices of engineering and biology. Its creations, bioartificial tissues and organs, are created from first principles to solve specific medical problems relating to repair, replacement, or regeneration of tissue or organ function. These are hybrid entities, both biological and artificial.

Synthetic "living" skin is a great example. The technique for creating it was first discovered in 1979 by Eugene Bell, professor emeritus of biology at the Massachusetts Institute of Technology and the founder of Organogenesis Inc., and was the first engineered body part to win approval by the U.S. Food and Drug Administration (FDA). Immediately following Bell's discovery was the inspired friendship and working relationship of chemical engineer Robert Langer and pediatric surgeon Jay Vacanti, who co-engineered the idea of biopolymer scaffolds for human tissue; the organ donor shortage problem was their motivation.

Since then, students of theirs and kindred colleagues – both in academia and from start-up biotech headquarters – have gone on to pursue the how-to's of engineering everything from bone to breast tissue, bladders, and even the heart. Steven Goldstein, biomedical engineering professor at the University of Michigan, directs his research efforts towards the complexity of bone regeneration, for example, and Michael V. Sefton, biomaterials professor at the University of Toronto, leads the ten-year initiative to grow a human heart.

Eugene Thacker
on biomedia

As computer science and molecular biology intermingle, how does our view of the human body change?
The integration of molecular biology and computer science is very interesting. When we think about computers or the Internet or digital this and virtual that, we think about immaterial things, which are completely mutable and portable and exist in a strange non-space. It's an abstract notion, yet the reality of computers is that they need hardware, cables, and infrastructure. When we think about biology, we think about the "stuff" of life, material, and things that are physical. It's a tangible notion, yet there are whole strands of biological thinking that go beyond the physical. When computing and biology come together, you get all sorts of strange hybrid artifacts, like an online genome database, or a DNA chip, or lab-grown tissues and organs.

In some instances, it means our notion of the body is becoming more immaterial or virtual. In other instances, it means the opposite: that, in fact, our notions of the biological and materiality are changing, and that biological materiality is being defined as informational. This means – rather than any kind of body anxiety or posthumanist fantasies of uploading your mind to a computer –, there is an insistence that we can control and manipulate biological matter through the lens of informatics.

Is the body itself a biotechnology?

Yeah, sure. I would say that it is, but with the caveat that it has to be articulated as a technology. I wouldn't say that this view is so all-encompassing that the body's mere existence means it's a machine or a technology. But once it's articulated or framed in such a way, then definitely it is a technology. That's what I would argue for a definition of biotechnology, that you enframe a "naturally occurring" biological process, and in doing that, you make it amenable to instrumentalization as well as to being used in all sorts of other contexts – and for ends that might not be biological at all, as with the field of DNA computing, which is a field of mathematics that uses DNA to perform computationally complex problems, whose application has nothing to do with biology; the application is mathematics.

How do you define biomedia?
Biomedia is a specific concept that's meant to describe the informatic reframing of biological components and processes. Packed within that are a couple of ideas. One is the framing or articulating of the biological, as I just mentioned. The other is the way that we articulate biology as a technology, through the lens of informatics, information, and information technologies. This is where we get our common notions of genetic code or the code script of life. But it's really through the lens of informatics and information technology that you get this combination of the immaterial and material, or biology and technology. It's about the process of identifying the biological, but looking at it through the lens of the informatic.

So we know the linear code script of life, but do we understand its processes and emergent behaviors enough to actually design life?
You and I could go online right now and download the entire human genome and all we'd really have is a long string of Ts, As, Cs, and Gs. The hardest part is yet to come, and scientists are certainly the first to recognize this. There are many basic, fundamental processes that happen in the cell – gene expression, cellular metabolism, cellular signaling – basic processes that you cannot look at in a reductive way. You have to take into account multifactorial, complex sorts of agencies in understanding them. It's not as if we have one gene that produces one protein and that one protein causes our eyes to be blue or brown or green. There are proteins that are made by DNA that go back to DNA and act on DNA in a circular process. The majority of phenotypic factors are polygenic, which means that there is more than one gene that's responsible for traits like eye color.

We should always be skeptical whenever we see a newspaper headline that boldly pronounces, "Scientists may have found the gene for X" – whatever X is. It's a very complex process and there's usually not just one gene that causes X to happen; there are usually multiple genes.

A number of researchers have been looking to different fields – complexity, self-organization, systems theory – in an effort to understand the genome in a more complex manner. Researchers like Stuart Kauffman have been saying this for decades and new fields like systems biology are trying to embrace this new approach. Whether this will translate into new drugs or therapy or maybe even a whole new paradigm in medicine is yet to be seen, but I think it's certainly an interesting step forward.

When computing and biology come together, you get all sorts of strange hybrid artifacts, like an online genome database, or a DNA chip, or lab-grown tissues and organs.

What are your thoughts on genetically modified foods and transgenics?
There's still a whole host of issues surrounding safety and regulation, but the thing that's interesting to me – and this is a logic that I think you can argue is in every biotech field – is the act of isolating some biological process and then reframing that process in a new context, in a context that's essentially like a factory. Transgenics is a great example. One of the most common uses of transgenics is for the production of particular molecules of compounds for drugs, which would then be employed for human use. There are goats, for example, that are genetically engineered to produce human insulin in their milk. This is where biology becomes the factory, an aspect of biotech that has totally changed the notion of what Marx called "living labor." [Labor] has been transformed to the effect that we begin to wonder what is performing the labor. Of course there are people working in labs, but there's also labor that's performed by microorganisms, enzymes and genes, 24/7, around the clock.

Breeding new species is not a new agricultural concept, but the new tools seem to render it new and scary.

You're exactly right. What is biotechnology? Really, it's just the use of "life." If you accept a really broad definition, there's archaeological evidence that in ancient Mesopotamia fermentation was being used, and certainly the breeding of livestock. So [in the broadest sense] biotechnology is almost concurrent with human civilization. Or, you can say that biotechnology is a particular relationship between human beings and their natural environment. Alternatively, you can be very specific in your definition of biotech and talk about a biotech industry, which emerged in the late 1970s with the first IPOs of the biotech companies, such as Cetus or Genentech, Inc. I think it's helpful to not limit oneself and to be mobile and travel between those two poles when talking about biotech, because it is a very big topic and very heterogeneous.

Do you think we'll be able to answer the question "What is life?" with microbiology?
Yes and no. When you talk about "life" to molecular biologists, it has traditionally meant biological life, "life" at the biological level. It hasn't necessarily meant social life, cultural life, psychological life. With molecular biologists, that has been a very specific question, but the question is also meant to evoke a sort of wonder at nature. This is why you see a lot of books with the title *What is Life?* It implies this notion that the concept of "life" will answer the big questions, but it will do it through a very specific lens. On the larger scale, I don't think a scientist would presume to – just because he or she studies the genome – know the larger existential questions about life. Unfortunately, when that has happened, we've gotten into trouble, from the dark side of the eugenics movement in the early part of the twentieth century up to now, with genetic discrimination. It's a very tough bridge to construct, between the molecular biology definition of life and our larger social, political, and cultural definitions of life. I think that's what the people in the humanities and social sciences can contribute, in trying to sort out the issues and in making that bridge.

Eugene Thacker is a biotech hobbyist and assistant professor at the Georgia Institute of Technology in Atlanta.

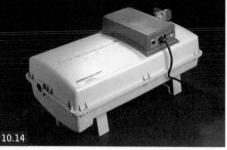

10.13 and **10.14** ULTRAVIOLET MAKES SAFE DRINKING WATER. Using UV-light technology, Ashok Gadgil's invention, UV Waterworks (10.14), disinfects four gallons of water every minute at a cost of about five cents for every thousand gallons. It is intended for a developing community of about 2,000 people, such as Bhupalpur, India (10.13). With one unit running 24/7, it would produce enough safe drinking water for the entire community.

10.15 BOTTLED SEWAGE. Since 1980, Zenon Environmental, Inc. has been perfecting membrane technology ("Zeeweed") for the purposes of overcoming global water shortages. A 2003 recipient of the prestigious Stockholm Industry Water Award, Zenon's magic membrane filters out microbes in drinking water, purifying even raw sewage. Newater is Singapore's answer to water self-sufficiency and a fine example of safe drinking water that was once regarded as waste.

10.16 and **10.17** TREATMENT FACILITY IN THE GUISE OF A GARDEN. Eco-designer John Todd (see page 186) installed a 600-meter canal restorer in an urban waste-filled waterway in southern China (10.16) and transformed it into a beautiful garden featuring over twenty species of native Chinese wetland plants (10.17). Wastewater, equivalent to that generated by 12,000 people, now enters the canal and, rather than sitting dormant and festering, supplies nutrients to the biological treatment components residing in the plant roots and fabric media.

Nature's lifeblood: It's too often assumed that our global water supply is without limit, yet available fresh water is less than half of one percent of the world's total water stock. We must look to the limits, clean what we have, and help those suffering from thirst.

Global consumption of water is doubling every 20 years, more than twice the rate of human population growth. If current trends persist, by 2025 the demand for fresh water is expected to rise to 56 percent above the amount that is currently available.

– Maude Barlow, National Chairperson, Council of Canadians

According to the United Nations, Europeans spend $11 billion per year on ice cream, $2 billion more than the estimated total money needed to provide clean water and safe sewers for the world's population. Astounding as that sounds, the reality is that billions go without clean water everyday and more than five million people, most of them children, die from illnesses caused by drinking poor-quality water every year.

According to Maude Barlow, in her comprehensive report on the world's global water supply ("Blue Gold"), before we can develop a worldwide water ethic, it's compulsory that we first acknowledge the profound human inequity in the access to fresh water sources around the world. Rather than insisting on the water-rich sharing with the water-poor, and unnecessarily damaging local bionetworks, we need to first look to sustainable solutions. In many cases, the technology to remediate this global condition lies dormant in labs and test facilities all over the developed world. In other cases, corporations like Oakville, Ontario–based Zenon Environmental, Inc., and water stewards like eco-designer John Todd and environmental physicist Ashok Gadgil recognize the capacity we have for action and refuse to sit still on the issue: Gadgil, based in San Francisco at the Lawrence Berkeley Lab, has made it his mission to globally distribute his invention, UV Waterworks, a lightweight, cost-effective unit that makes dirty water safe to drink by way of ultraviolet light. John Todd, founder of Ocean Arks International, designs and builds "living" systems to restore balance to distressed ecosystems.

Ashok Gadgil
on safe drinking water

What happened in northeastern India in the summer of 1992 and how did it affect the direction you took with your career?

There was an outbreak of a mutant strain of cholera in Bengal, which became known as "Bengal cholera." Because the surface protein on this mutant strain was slightly different from what's common, all of the cholera vaccines were ineffective in protecting populations from this particular strain. So thousands of people contracted cholera within weeks. In one month alone, as the epidemic spread, some 10,000 people died from this cholera epidemic, in a single state in India. Soon after, this particular strain spread from India to Bangladesh and also turned up in Thailand. That's when I decided to do something about it. I had to do something about it because I had been thinking about ultraviolet (UV) disinfection for quite some time as a potential way to disinfect drinking water inexpensively for poor communities in poor countries.

What is the scale of the global drinking water problem right now?

About two billion people, roughly one third of the global population, need to go outside their home to fetch water for daily use. Of those, 1.2 billion people don't have access to safe drinking water; they are forced to rely on biologically contaminated water, in most cases. This leads to a large number of diseases and deaths, particularly for children below age five. Young children have low

resistance to dehydration, which is the resultant condition of diarrheal diseases.

With all of the advances in public health, technology, and medicine today, why is it that still 20% of our world's population is without access to safe drinking water?

Right, it is 20% if you exclude the residents of large metropolitan areas like Jakarta, Bombay, Cairo, and Mexico City; if you include them, then the number rises to about 30%. We already have the science and technology to address this problem. It is not anymore a scientifically inaccessible domain, intellectually. It's something that we can do. It just hasn't been done, for a variety of reasons, including inadequate investment in water infrastructure. There is also the mindset that in the developing countries we'll just follow the model of what's been done in the industrial countries, which is to pipe pressurized safe water 24/7 to everybody – and that requires a level of investment and a level of water availability that's often just not supportable. There is also the problem of governance. In many developing countries, the political will to provide safe drinking water dissipates as soon as the politically most vocal and powerful segments of society have access to safe drinking water. Sadly, those who are relatively voiceless and politically weak are left to fend for themselves.

Is water a fundamental human right or is it a commodity?

This is a societal question and the views are sharply divided. Initially, the overall opinion of global policy-makers was that water should be considered a right. However, it appeared after some 20 years of experience that even if they considered it a fundamental human right, investment and aid kept on getting funnelled into supplying water only to those who had political access or a political voice in the developing world. It also led to huge incompetence and inefficiency in the supply management of water systems in the developing countries. This was because of enormously bloated bureaucracies with no performance metrics and no accountability. About three years ago there was, after much pushing and debate, a sea change in the way this problem was viewed, at least in the industrial countries. Now the viewpoint, led by the World Bank, is that water should be treated as a commodity, and full cost should be recovered for its supply, which would then, one hopes, amount to some kind of financial accountability in terms of supply costs, and presumably lead to improved efficiency.

How could a strategy based on public-private partnerships work?

This type of strategy draws on the best of both the private sector and the nongovernmental organizations' grassroots outreach efforts. Relying on government alone has not been successful. We have not been able to get safe drinking water to people who need it badly, with a horrendous toll: About 400 children die from diarrheal diseases per hour in the developing countries – every one of the deaths preventable. The public-private partnership vision is not easy to implement, but when it works, it works beautifully. You have elements of the private sector – the flexibility, the dynamism, the entrepreneurial spirit, and the ability to rapidly expand services and go out and do something – coupled with the sense of public purpose to do what's considered essential.

have been around for a long time. To produce the UV light, one essentially passes an electric arc through mercury plasma, which causes the mercury atoms to excite and de-excite. The lamp in the UV Waterworks units is made of quartz, so the UV light can pass through. Normal glass will block it. As well, the UV lamp doesn't have a layer of phosphor on the inside of the glass tube so that you don't get visible light; you just get UV light coming right out. This is a very efficient process.

In terms of energy use, 60 watts of electrical power – which is comparable to the power used in one ordinary table lamp – is enough to disinfect water at the rate of one ton per hour, or fifteen litres per minute, which is approximately two and a half times the water flow in a standard bathtub faucet. This much water is enough to meet the drinking water needs of a community of 2,000 people.

The arsenic crisis in Bangladesh is rightly described as the largest case of mass poisoning in the history of mankind. Forty-six million people are forced to choose between arsenic-laced groundwater or biologically contaminated surface water.

Where and how many of your UV Waterworks units are distributed around the world?

Over 300 units are functioning daily throughout the developing countries. There is a handful in the U.S., but most of them are in the developing countries; and most of those – say, 200 out of 300 – are in Mexico and in the Philippines. The rest are scattered throughout Asia, Africa, and Central America, all the way from South Africa to some in Nicaragua and Honduras. There are also some in India, Nepal, and Bangladesh.

What do these units do to pathogenic bacteria and viruses?

The light that's used in the UV Waterworks is known as C band, the short wavelength end of the UV spectrum. UV light causes the adjacent base pair in the DNA helix to fuse together, so that when the DNA tries to replicate, or the organism tries to replicate, it cannot unzip the DNA, and it dies. UV disinfectors, based on this principal,

What is your next big project?

I am currently working on trying to figure out a way to remove arsenic from drinking water in Bangladesh. The arsenic crisis in Bangladesh is rightly described as the largest case of mass poisoning in the history of mankind. Forty-six million people are forced to choose between arsenic-laced groundwater or biologically contaminated surface water. Given that choice – either slow death by arsenic poisoning or immediate extreme sickness with surface water pathogens – they inevitably and consistently choose the arsenic-laced groundwater. For the arsenic removal process, the goals are similar to UV Waterworks. In terms of cost, it must be affordable; it must be cheaper than what's being tried today in Bangladesh. It must be highly effective. It must have very large margins of safety. And it must be very easy to use. We have some good, exciting preliminary results in hand that suggest that we might be able to meet all these goals.

Ashok Gadgil is an environmental physicist at the Lawrence Berkeley National Laboratory in Berkeley, California.

We will eradicate poverty.

0.9

USA

France

Germany

Argentina

Italy

Russia
Chile
Japan

0.6

Mexico

0.4

Brazil

0.2

India

0

1870   1913   1950   1995

The convergence of the Human Development Index over time, 1875–1995.

New systems of design – of communication, production, evolution, and exchange – have the potential to create shared wealth on an order of magnitude the world has never seen. Design and its capacities promise to make this century a new era of wealth worldwide. For nearly 200 years there has been a steady decline in the percentage of the world population living on less than a dollar a day, from 85% in 1820 to less than 5% today. Over the past 20 years, despite an increase in world population of 1.6 billion, for the first time in history the number living in poverty has dropped by 200 million.

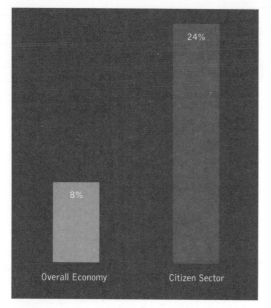

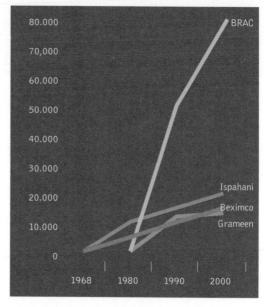

11.01 Of the total non-agricultural employment growth from 1990 to 1995 in the combined economies of the U.S., Japan, and the EU, the citizen sector (social entrepreneurialism) grew three times as much as the rest of the economy.

11.02 Over the past two decades, the number of workers in citizen groups in Bangladesh (BRAC, Grameen) grew to four times that of workers in the top Bangladeshi business groups (Beximco, Ispahani Group).

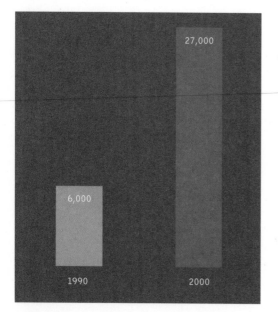

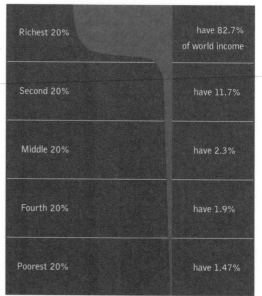

11.03 In just one decade, the number of citizen groups operating internationally increased from 6,000 to a whopping 27,000. The global reach of social entrepreneurialism is unstoppable.

11.04 The world's richest 50% (the top half of this graph) have 95% of the world's income. The poorest 50% (the bottom half of this graph) have what's left, just 5%. Conceptualizing the world's richest and poorest this way makes evident the sort of gross imbalance that is at the root of social revolution.

The citizen revolution: From all corners of the globe, we are now bearing witness to an emergence of social entrepreneurs with ethics as powerful as their conviction to do the greatest good for all.

A global economic system is developing without a global society to provide the necessary framework for these developments or to correct the distortions brought about by market fundamentalism.

– George Soros, philanthropist, *The Crisis of Global Capitalism*, 1998

Citizen groups, social entrepreneurs, nongovernmental organizations (NGOs), non-profit associations. Each one under a different moniker, all of them ride the same wave that washes over the public and private sectors of global wealth and politics. Distributed geographically but connected in their common motivation to represent the voices not accounted for otherwise, these groups embody in aggregate a movement toward the sort of global civic society that philanthropist George Soros believes we need to counter the forces of free-market fundamentalism and economic globalization.

Some numbers: Indonesia saw an increase from 1 to over 2,000 environmental groups from 1983 to 1997; Slovakia from 10 to over 10,200 citizen groups in a ten-year period (1989–1999). In twenty years (1980–2000), Brazil increased its citizen groups from under 5,000 to over 1,000,000 (400,000 of which are officially registered); and the U.S. has grown its IRS-registered citizen group numbers from 464,000 in 1989 to 1,100,000 in 2002. The citizen sector job growth was three times that of the rest of the economy in the U.S., European Union, and Japan from 1990 to 1995.

According to the international nonprofit organization Ashoka, the number of international citizen groups was 6,000 in 1990; by 2000, the count was up to 27,000. Former McKinsey & Co. financial consultant Bill Drayton, founder of Ashoka, recognizes this pattern-changing force and cultivates it by supporting social entrepreneurs with big ideas. Through rigorous selection criteria based on ethical fiber, creativity, and entrepreneurial skills, Drayton and his associates have acquired a base of Ashoka Fellows, "changemakers" who see their novel ideas through to adoption by independent organizations, and eventual national policy change.

Bill Drayton
on the citizen sector

Describe the transformation of the citizen sector that we're witnessing today.
First, it helps to look at the historical framework. Starting around 1700, the business sector went through a transformation, one that empowered anyone with an idea to start a business. This shift was so effective that, over the past three centuries, it compounded productivity in the business half of society two to three percent a year. An equal shift in the social half of the world did not happen. As a result, societies became half-stunted, backward, and relatively unproductive, while the business sector grew dramatically. The very recent citizen sector breakthrough is a direct result of this intolerable imbalance.

The social entrepreneurial movement started earlier, of course, with individuals like Florence Nightingale and Maria Montessori, who were as brilliant, in a social context, as Carnegie or Rockefeller were in business. Despite these remarkable pioneers, however, the social sector as a whole did not make the jump to the entre-preneurial/competitive architecture that had allowed business productivity to soar.

Roughly two and a half decades ago, the social sector as a whole began the process of tipping from premodern to the same entrepreneurial/competitive architecture adopted by business centuries earlier. Control by a few percent was no longer cutting it in a world of ever more pervasive and rapid change. Social entrepreneurs have led this transformation. However, two decades ago we

didn't even have the word "social entrepreneur"; when we started talking about it, people would go glassy-eyed, and the really smart ones would say it was an oxymoron.

What sort of character becomes a social entrepreneur?
The core psychology of a social entrepreneur is someone who cannot come to rest, in a very deep sense, until he or she has changed the pattern in an area of social concern all across society. Social entrepreneurs are married to a vision of, for example, a better way of helping young people grow up or of delivering global healthcare. They simply will not stop because they cannot be happy until their vision becomes the new pattern. They will persist for decades. And they are as realistic as they are visionary. As a result, they are very good listeners. They have to hear if something isn't working; and, whenever they do, they just keep changing the idea and/or the environment until their idea works. They are intensely con-cerned with the how-tos: How do I get from here to there? How do I solve this problem? How do these pieces fit together?

In the eyes of Ashoka, is the citizen group the same thing as the NGO?
We cringe whenever anyone uses the term NGO, or nongovernmental organization – or nonprofit, for that matter. You can't define a sector by what it isn't. Again, the history is interesting: the Europeans saw something new and they said, "Oh, it's a nongovernment organization." The Americans saw something new that was not what they expected and called it a "nonprofit." (A brothel, for example, is usually a nongovern-mental organization!) So we prefer to focus in on the active ingredient, the citizen individually or in a group, who takes the initiative in an area of public concern, be it to provide a service or introduce a needed change.

How are social entrepreneurs affecting global politics?
I believe that social entrepreneurs are the cutting edge of the democratic revolution. They and the groups they lead exercise enormous power. With an idea, they change the whole system. They have no armies. They cannot force people; theirs is the power of caring and persuasion.

The organizations that all these changemakers and social entrepreneurs are building also link the individual citizen to the government, from a posi-tion of power. If you care about the environment – any aspect of the environment – you can choose one or more environmental groups that truly represent your interests. In many different ways – the individual citizen, empowering everyone to be a changemaker, the creation of new patterns that make the whole system work better – social entrepreneurs are at the heart of democracy in action.

What is the global collaboration challenge that you encourage?

One, how can you solve the world's problems if we don't work together on them at the global level? There is no way that any nation can solve, say, the world's environment problems only at the national level. Not China, not the U.S., no matter how big and powerful we are. How can we build a global financial system that's safe if we don't build it at the global level? We are at a stage in the evolution of our planet that we have to think and work together.

Second, in any case, there's huge efficiency and satisfaction in our working together as a field. Worldwide, when you take all of the partial answers from individual countries and put them together, we can see what's universal. And then we can work together to spread those universal insights. Working together is many times more effective than the sum of our individual efforts.

At present, we have an antique, stuck, and client-deaf financial services structure serving a new dramatically transformed citizen sector. We desperately need a new, diverse, client-oriented financial services "industry" to serve the social sector. How are we going to get from here to there if we don't work together to encourage entrepreneurship in social investing and to encourage new people to enter the field? There is no issue that drives Ashoka's social entrepreneur members more crazy than this, which makes it front and center in our agenda.

By way of jujitsu leverage points!

Exactly. Every entrepreneur, business or social, succeeds because he or she knows where the jujitsu leverage point is and presses toward it with every ounce of skill and energy. Finding this magical leverage point is the heart of the matter. The most common jujitsu point comes from demonstrating just how attractive a new idea is and then, through deft marketing, setting off a chain reaction of others rushing to capture these advantages.

The entrepreneur may have to experiment for years to find how to get to the jujitsu point, and he/she must know surely where it is. For example, Ashoka is well into the work of demonstrating that trading "hybrid business/social value-added chains" is a key step in overcoming the many stupidities caused by the several centuries of divergence between the business and social halves of society. By actually putting together new production and distribution chains that draw on the unique strengths of business and social institutions at different points in the production system, we can demonstrate enormous new markets and profit potential for business, very major new revenue flows for the citizen groups, and better-priced products and services for everyone. Once we have demonstrated this for slum dwellers and

> ## How can we build a global financial system that's safe if we don't build it at the global level? We are at a stage in the evolution of our planet that we have to think and work together.

building product companies, small farmers and piping companies, forest dwellers and forestry companies, and health care recipients and providers, the financial pages and business schools and management-consulting firms will be all over the idea. Not to mention the business and social competitors of the pioneers! It is precisely this sort of self-multiplication of an idea that is every entrepreneur's dream.

How do budding social entrepreneurs hook up with Ashoka?

www. Ashoka.org. It's all there. There are also offices in some fifty countries around the world. A career in social entrepreneurship is quite magical. It offers huge impacts, a direct fit with your values, increasing public recognition and support, no glass ceilings, and even – for the first time in centuries – salaries that are growing relative to business (as we catch up in productivity). Amidst the present soaring demand and lagging perception, the demand far outstrips the supply.

Bill Drayton is the founder of Ashoka: Innovators for the Public, a global nonprofit organization that invests in social entrepreneurs around the world.

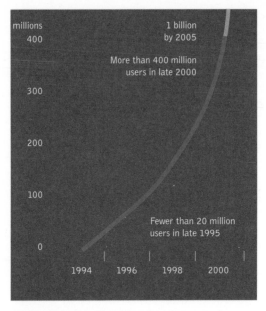

11.05 Since the mid-1990s, when there were fewer than 20 million Internet users, more and more people have learned how to surf. At the end of 2000, there were 400 million users. By 2005, it's estimated that one billion people will have Internet access.

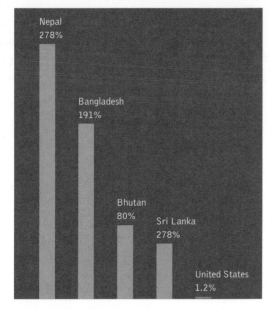

11.06 The monthly Internet access charge as a percentage of average monthly income is the real cost of being connected. Nepal: 278%; Bangladesh: 191%; Bhutan: 80%; Sri Lanka: 60%; and the U.S.: only 1.2%.

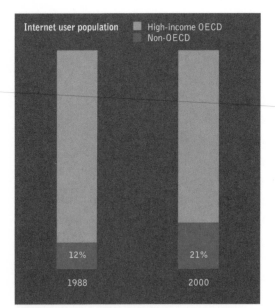

11.07 The digital divide begins to narrow, ever so slightly. High-income countries that are members of the Organization for Economic Cooperation and Development (OECD) (countries that have only 14% of the world's people) represented 88% of the Internet user population in 1998 and 79% in 2000. Put another way, the percentage of non-OECD countries online was 12% in 1998 and grew to 21% by 2000.

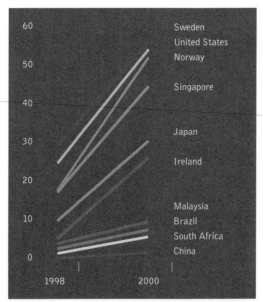

11.08 Internet users have increased globally. In just two years, as a percentage of national population, the numbers grew from 30% to 50% in Sweden; 26% to 54% in the U.S.; 29% to 52% in Norway; 29% to 45% in Singapore; 10% to 30% in Japan; 5% to 28% in Ireland; 4% to 8% in Malaysia; 3% to 7% in Brazil; 2% to 6% in South Africa; and 1% to 3% in China.

The digital divide: Provide access to all global knowledge systems. Developed and developing worlds alike will use the tools, become empowered, and get connected to the global network.

The fundamental cure for poverty is not money but knowledge.

– British economist Sir Arthur Lewis (1915–1991)

Known as "capacity gurus," the United Nations Development Program (UNDP) is the development arm of the United Nations that runs technical assistance offices around the world, in 134 developing countries, providing broad assistance in six core areas: poverty reduction, democratic governance, environment and energy, response to HIV/AIDS, conflict prevention and resolution (in some countries), and Information Communications Technology (ICT) for development. The ICT for development priority seeks to bridge the digital divide and help countries develop what UNDP ICT Director Stephen Browne calls their "road map" for the information society. One initiative of several is their Radios for the Consolidation of Peace project with the Freeplay Foundation, where small, illicit arms are accepted in exchange for windup radios.

Outside of the UNDP ICT, but (in principle) supported by it, is an extraordinary development from Bangladesh, a country that for so long has been overcome by isolation and impoverishment. Grameen Bank founder Mohamed Yunis and teammates have put mobile telephones in the hands of poor, remote, and largely illiterate village women to astounding effect. The women take out micro-loans from Grameen Bank to purchase the oversized handsets and establish themselves as mobile local operators. For a few *taka* a day they hire the phones out to whomever wants to make calls for whatever reason, and as a result, these women can earn between US $50 and $500 per month, in a country where the majority of the population lives on barely a dollar per day. A bonus feature: All Grameen phones come equipped with direct-dial to the prime minister. This provides not only political empowerment to these women, but also self-empowerment, which is just as important.

Stephen Browne
on the digital divide

When it comes to poverty reduction, what role does Information and Communications Technology (ICT) play?

There are innumerable ways in which ICT can be applied to deliver better services to poor people, whether it's telemedicine, using ICT to bring medical care to people in remote areas, or e-schools, using ICT to enhance the delivery of curricula to remote areas. There is a wide variety of practical applications, and I'm referring to actual examples in a growing number of countries where these applications have been put into practice with great effect. The most important general contribution that the emerging information society can provide to people everywhere, however, including the poor, is access to knowledge and an empowerment mechanism by which they can themselves take hold of their own lives and seek to improve them.

In January 2004, the United Nations Development Program (UNDP) and Microsoft jointly put out a press release announcing a partnership. What has developed as a result?

This is something that will build over time. I think that the synergy between UNDP ICT and Microsoft results from our respective global vocations, but unlike Microsoft we look for highly differentiated solutions to problems at the country level and within countries at the local level. In other words, we eschew the idea of there being a single solution for every problem. Now, you might say that is contrary to the Microsoft idea, which is complete uni-

formity in their software applications. But no, we're helping Microsoft identify areas in which their standard software applications can actually be applied in specific organizational or other contexts. I think that we've learned a lot from each other so far. Concretely, we are building this partnership in a few countries, such as Morocco and Mozambique, and we hope to spread this outward.

We have certainly not decided that all of the software solutions for the developing countries are going to be based on Microsoft technology. We are very keen to encourage countries and organizations and individuals within them to make their own choices of the best software solutions. In Afghanistan, for example, we identified telecenters that could be empowered with free Microsoft software, while at the same time educating people about the advantages in other contexts of open-source technologies. Where we can work with Microsoft and see that their product can bring an advantage to a particular situation, then we'd be delighted to help them find an opportunity. On the other hand, we look for other solutions, many of which include the rivals of Microsoft. Chief among them is the open-source nonproprietary software.

As with countries like Bulgaria and Brazil, which tend to favor free and open-source software, right?

Yes, I was about to mention Bulgaria as a good example of a country which, for its own reasons, has decided that in much of its public sector it wants to go for an open-source solution. And we've been ready to help them on that. As I've said, we try to be non-conflictual about it. We believe in using partnerships to the advantage of our clients. And this doesn't mean that we have a single standard solution. That doesn't mean to say that we are going to be inviting all of our program countries around the world to use Microsoft exclusively just because we have a global arrangement with them. We will be fairly opportunistic and see where it is that countries find Microsoft to be an advantage and where they don't.

If and when there is a greater global demand for free and open-source software (over proprietary), I wonder if Microsoft will respond accordingly?

That's a good question. We sometimes use an analogy with HIV/AIDS treatment, which has become, not nonproprietary exactly but generic, available at drastically reduced cost to the developing countries. We would like to see the same sort of thing happen here. And I think it is beginning to happen through Microsoft, providing what are otherwise to some of us quite expensive products at drastically reduced prices, or even providing them free of charge. We hope that this could be one of the great benefits of the partnership.

How are citizens of the developing world beginning to use the Internet to promote political accountability?

This is where I believe our work on ICT for development as one priority and the work on democratic governance as a second priority very definitely overlap. I spent some of my most interesting years with the UNDP as the UN representative in Ukraine very shortly after it had become an inde-

extraordinary figure. That's one billion out of about seven billion people. When you think that only a few years ago, in the year 2000, there were only 200 million users of the Internet, it's gone up by five times in five years, more or less. That's pretty extraordinary. So, to a considerable degree, we are becoming a partially networked world.

But there are two major impediments to a fully networked world. One is the infrastructure, and how we can reach the most remote areas in a way that is financially viable. It's a big challenge, although there are a number of solutions now emerging. The other problem is language. Although the proportion of pages of the Internet in English is steadily diminishing – and I think soon the next most popular language is going to be Chinese – there are inhibitions to people communicating across countries in different languages; but even more important, of course, is literacy rates. People who are not literate are not able, of course, to take the same advantage of being connected as those that are.

> The digital divide is really only a technological manifestation of the same old development divide with which we have been familiar for so long. We have to tackle both.

pendent country. I was able to appreciate there the idea of the Internet having contributed to the disintegration of the Soviet Union. I experienced the post-Soviet phase, and saw the extraordinary impact that access to information had for 50 million people who, with very few exceptions, had really never had access to any kind of information from the outside, or even a dialogue through international telephone lines. We were providing them with free dial-up service, and so I witnessed some Ukrainians who were ready to sit in front of a computer for almost 24 hours without stopping, notwithstanding the language barrier – as the Internet was very much and still is somewhat dominated by English – just browsing and informing themselves and communicating with people who had been remote to them before.

All of that can certainly have an important impact on political processes. I don't believe that the mere existence of the Internet is an automatic panacea for closed societies. But there's no question that it's going to be more and more difficult for societies to remain enclosed and immune from opinions within and outside of their country.

What will it take to get the entire world population participating in the global networked economy?

I was looking at some statistics the other day. There are expected to be by the end of this year one billion users of the Internet. That's a pretty

The digital divide is really only a technological manifestation of the same old development divide with which we have been familiar for so long. We have to tackle both. It's not merely a question of miraculously delivering telephones to every remote corner of the developing world. We will overcome the digital divide if we overcome the societal divide. This means bringing everybody, without exception, fully into the development process.

Do you feel that the bridges are being built?

Yes, I do. It partly depends on the activities of organizations like UNDP and other development partners, like The International Development Research Center (IDRC) and the Canadian International Development Agency (CIDA). It largely depends on the kinds of initiatives which countries themselves can take. There's a huge amount that developing countries can do to open themselves to the information society. There are large and quite complex political decisions that they have to make. More openness to investment by technology companies in these countries will help to facilitate the kind of change which I think could bring about what is being described as a revolution.

Stephen Browne is the director of the United Nations Information Communications Technology for Development special initiative.

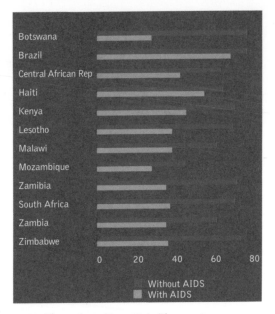

	Female	Male
Sub-Saharan Africa	5,700,000	2,800,000
South and Southeast Asia	93,000	590,000
Latin America	17,000	260,000
North Africa and Near East	110,000	41,000
East Asia and Pacific	87,000	200,000
Eastern Europe and Central Asia	85,000	340,000
Caribbean	72,000	59,000
North America	47,000	100,000
Western Europe	33,000	55,000

Female
Male

Botswana
Brazil
Central African Rep
Haiti
Kenya
Lesotho
Malawi
Mozambique
Zamibia
South Africa
Zambia
Zimbabwe

0   20   40   60   80

Without AIDS
With AIDS

11.09 Out of the 11,745,000 young adults ages 15 to 24 living with HIV/AIDS around the world, 7,300,000 are female. Adolescent men outnumber women in all global regions except Africa.
In Sub-Saharan Africa, 5,700,000 young women are infected with HIV/AIDS, more than double the number of young men. The same condition exists in North Africa and the Near East.

11.10 The projected impact on life expectancy (to 2010) at birth, in years, of those living with AIDS is dramatic – in Africa, especially. The numbers are: Botswana (27); Brazil (66); Central African Republic (41); Haiti (53); Kenya (44); Lesotho (37); Malawi (37); Mozambique (27); Namibia (34); South Africa (36); Zambia (34); and Zimbabwe (35).

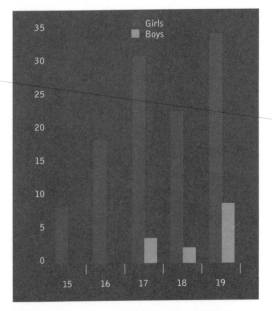

Girls
Boys

35
30
25
20
15
10
5
0

15   16   17   18   19

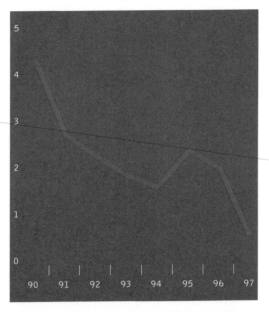

5
4
3
2
1
0

90   91   92   93   94   95   96   97

11.11 Research sponsored by the Bill & Melinda Gates Foundation reveals the gender disparity in HIV prevalence rate among teenagers in Kisumu, Kenya: at age 15 (no boys and 8.3% girls); at age 16 (no boys and 17.9% girls); at age 17 (3.6% boys and 29.4% girls); at age 18 (2.2% boys and 22% girls); and at age 19 (8.6% boys and 33.3% girls).

11.12 According to the World Health Organization, Uganda's overall governing body and its Ministry of Health, in particular, did wonders in the face of HIV/AIDS. The HIV prevalence rate among 13- to 19-year-olds in Masaka, rural Uganda, for example, has dropped from over 4% in 1990 to less than 1% in 1997.

The gender power imbalance: Education and economic opportunity are key to breaking women and girls out of the vicious cycle of inequality, poverty and, increasingly, HIV/AIDS.

Women in poor countries are bearing a disproportionate share of the global AIDS epidemic and need political empowerment as much as medicine to fight it.

– Nafis Sadik, UN special envoy for HIV/AIDS in Asia and the Pacific

In his address to the International Woman's Health Coalition (IWHC) in January 2004, UN Secretary General Kofi Annan said, "Study after study has taught us that there is no tool for development more effective than the education of girls and the empowerment of women. No other policy is as likely to raise economic productivity, lower infant and maternal mortality, or improve nutrition and promote health, including the prevention of HIV/AIDS."

Dr. Nancy Padian, executive director of the Women's Global Health Imperative (WGHI), sets up studies on HIV/AIDS and reproductive health around the world, with the purpose to empower young girls and women. All WGHI studies involve education, economic development, female-controlled contraception, vocational training, and – in the case of the organization Shaping the Health of Adolescents in Zimbabwe (SHAZ) – mentorships with local businesswomen. The thrust of SHAZ is to investigate the correlation between economic independence and sexual relations between young girls and "sugar daddies." One female subject told WGHI researchers: "If you refuse, you stay poor. If you take his money and refuse sex, he will rape you." Sugar daddies are often HIV-positive.

There is hope, though, when we look to Uganda, a country that managed to turn the corner on the AIDS pandemic. In 1993, Uganda had the highest HIV infection rates in the world. A swift government response that involved collaboration with international organizations and NGOs contained the escalating spread of HIV.

Nancy Padian
on women's health

Describe the sense of urgency you feel right now with respect to the HIV/AIDS crisis in Zimbabwe.
It's very hard to describe the magnitude of the epidemic in a country like Zimbabwe unless you've been there. It's staggering. Everywhere you look, there it is. There are cottage industries that build coffins. There are orphan-feeding programs virtually everywhere. If you go into the hospitals, there's complete overcrowding. There's no one that I work with there who hasn't been touched by it somehow. It's completely common that when someone's not around, they're attending a funeral. About one-third of the population is infected, and it cuts across all socioeconomic levels.

What is your main objective with the Women's Global Health Imperative (WGHI)?
To deal with HIV/AIDS and other sexually transmitted infections and even unintended pregnancy in the context of addressing our overarching theme: gender disparities and gender inequities, and how that gender imbalance places women at greater risk for HIV, other sexually transmitted infections, and unintended pregnancy. We have two major areas of research right now. The first is exploring women-controlled methods of prevention of these outcomes. Male condoms, when used correctly and consistently, are the most effective way to prevent HIV transmission heterosexually. The problem with male condoms is that men control their use and, insofar as a woman would have control over when her male partner uses condoms, it

requires negotiation and willingness on his part. So we're looking at methods that women might be able to control, such as microbicides and the diaphragm – methods that women can use themselves without having to rely on negotiating with their male partner. The other way we're looking at gender and vulnerability to HIV is by way of economic intervention. Our hypothesis is that if you can provide young women with economic opportunity and make them economically independent, then they will be less reliant on sexual partners, particularly older sexual partners, for material goods and, in some cases, even survival.

Who is part of the WGHI network?
In Zimbabwe, our major collaborators are the University of Zimbabwe and the medical school. In India, our major collaborator is Samuha, a nongovernmental organization. In Mexico our major collaborator is the National Institute for Public Health, in Cuernavaca. In the U.S., we are an institution based in San Francisco, and we collaborate with many community groups. Most of our work here is done in what's called the Mission District in San Francisco, where there are a lot of Latino immigrants.

Tell me about the HIV Prevention Trials Network (HPTN).
The HPTN is a large program funded by the National Institutes of Health in the U.S. Essentially, the way the network works is that you apply to be a member of the network – and the reason why such a network is critical is because most prevention studies require huge sample sizes. In order to assess whether the trial that you're working on was effective and really prevented HIV, you need very large numbers of people; you need to see that enough HIV infection was prevented. The HPTN is a network because it requires more than one country, more than one site to participate in a prevention trial. It's a very effective means of launching prevention trials, and that's how we're doing our microbicide studies. In my diaphragm study, we had to create such networks on our own.

Medical ethicist Solomon Benatar once said, "Saving lives in poor countries almost never results predominantly from costly, novel medical research. Existing options must be explored as part of our HIV prevention strategy." Talk about the diaphragm in this context.
After much examination, biologically, we have reason to believe that the diaphragm, a readily available technology, can protect against HIV in a way similar to the way it protects against pregnancy.

We believe, at the very least, that it's worth testing. So as we are simultaneously looking for new strategies, new microbicides, we are testing this promising method that already exists. There are so many reasons why the diaphragm is very appropriate to the locality. Number one, it's an existing option that we can get in the hands of many women in Zimbabwe, and it allows for clandestine use, which is so important where there's this gender power imbalance, and resistance on the part of men to using male condoms.

But testing the diaphragm has its challenges. I think people inherently have a more difficult time believing that you can have a new use for something that's been around for a long time versus something brand new. Convincing people that the diaphragm might actually be a strategy that would work for HIV prevention is very difficult.

story where new infection rates have been reduced. One of the hallmarks of Uganda's program is that it implemented multicenter involvement in not only dealing with HIV but also in promoting women's rights and giving women a strong voice, not only in terms of HIV prevention but in other aspects as well: having a voice in parliament and being involved in decision-making positions. AIDS has allowed us to focus on these gender power imbalances. Now is the galvanizing moment in which, clearly, we have to make a change. Empowering women will have an effect not only on their vulnerability to HIV, but also on all other health outcomes and, by extension, on the countries themselves. There can be nothing but good that comes out of this. If this is what it took to get us here, then maybe that's a collateral benefit of this epidemic.

Now is the galvanizing moment in which, clearly, we have to make a change.

Do you think the solution is to not rely entirely on political leadership but more on philanthropy, like the Bill & Melinda Gates Foundation?
I have to believe this because I work in Zimbabwe. If I thought that the only effective solution would come from political involvement, I would be working in the wrong country. The Bill & Melinda Gates Foundation has been incredible in really filling a niche both in terms of resources and willingness to fund my study and other similar studies where the priority is placed on the kinds of things Benatar was referring to – projects that have a high likelihood of being sustainable. I think the foundation has been absolutely at the forefront of that.

If you believe that change will occur only with governmental support, then what does that mean for countries where the government's in turmoil? Do you not work there? Do those people then have to suffer as a result? If anything, the need is greater in these places. There are religious leaders and leaders at other levels, and you have to believe that the network that you're putting together will be able to maintain itself once good, strong, stable leadership is in place.

This strengthens the point about empowering women worldwide in a healthcare context.
Yes, and I'm really glad you brought it up because that is the theme of our research. In terms of changing social norms, Uganda is a great success

Are countries such as India, Asia, and areas in Eastern Europe now at the tipping point with HIV/AIDS?
I think rather than debate whether India's going to be the next Sub-Saharan Africa, we need to acknowledge that a huge number of people are infected there. If you look at the rate of infection over the population, yes, it looks like it's at the beginning of the epidemic. So let's get in there, do something now, and not even answer, "Is it going to be like Africa?" We must prevent that from even being a possibility.

As a global community, what would it take to be able to say that we did everything in our collective power to provide some sort of solution?
At the University of California at San Francisco (UCSF), we're developing a coalition of everyone who's working on HIV in the international scene, so that we work together as one larger organization and focus on international health. I'm hoping that there will be more willingness on the part of researchers, activists, and community-based people to work together as one coalition, too. I know it sounds a bit like "it's a small world," but I think that if we can all work toward this in our own line of work, we will change the social norm of research. I hope too that this coalition will equally involve resource-rich countries and resource-poor countries, working together.

Nancy Padian is the executive director of the Women's Global Health Imperative based at the University of California San Francisco.

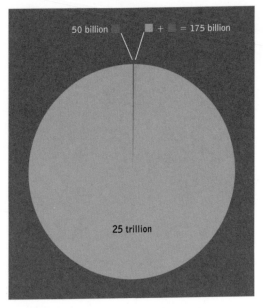

11.13 The U.S., Canada, Western Europe, Japan, Australia, and New Zealand promised to give 0.7% of their total GNP ($25 trillion) to the world's poor. So far, $50 billion has been delivered.

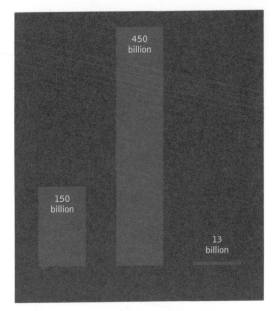

11.14 In the U.S., the Clinton Administration allocated $150 billion to the military. The Bush Administration spent three times that. How much did the Bush Administration direct to the fight against global poverty? A meager $13 billion.

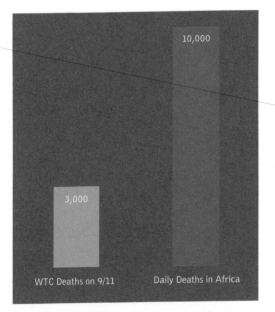

11.15 On September 11, 2001, in New York City, approximately 3,000 people died in the terrorist attack on the World Trade Center. Everyday in Africa, at least 10,000 people die from AIDS, tuberculosis, and malaria.

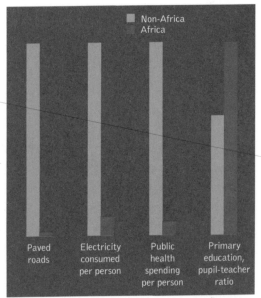

11.16 The non–African developing world is suffering. Africa is dying. In close analysis of averages in infrastructure, education, and public health spending from six African countries (Ethiopia, Ghana, Kenya, Senegal, Tanzania, and Uganda) versus elsewhere in the developing world, the discrepancies are staggering.

Global poverty: Before we can inspire global citizens to fight the war on poverty, we must cut through all rhetoric on the "war on terror." It's time to address the true terror that undercuts global security: insufficient aid for development.

It is not crazy for us to think about having within our power, uniquely for the first time in the history of the world, the chance to end extreme poverty within a generation. That is what the numbers show.

– Jeffrey Sachs, director of the Earth Institute

When citizens of the one world we share – in Africa's heartland, most poignantly – are starving and dying of curable diseases every day, it is our duty to direct our dollars to sustainable economic development, not already bloated military budgets. An integrated strategy will result in the deepening of global security and the alleviation of abject poverty and its indicators: violence, terror, and disease. It will depend on the formation of policies for the prevention of future conflict and partnerships between the world's rich and the world's poor. Countries in need can't do this alone.

The millennium development goals put in place by all UN member states in 2000 to reduce extreme poverty by 2015 required that poor countries pursue good governance and responsible economic and social stewardship, while rich countries helped "well-governed poor countries through expanded aid, trade, and technology transfer." Many African countries – Ghana, Senegal, Mali, Benin, Ethiopia, Uganda – have shown exceptional leadership and effort in the transformation of dire political scenarios into thriving democracies. They have the will, but they still lack the way.

Although the governance is admirably in place in many instances – in Africa and elsewhere – the means are sorely lacking to build the necessary infrastructure and social services that will help impoverished nations on the road to self-sustainability and eventual prosperity.

Now that we *can* create a world of shared prosperity, what will we do?

Jeffrey Sachs
on poverty reduction

Since the UN Millennium Development Goals were put forth as a global challenge, how well have we rallied together to meet the needs of the world's poor?
These goals were set in September 2000 at the Millennium Assembly of the United Nations. In fact, most of them were recycled from commitments at international gatherings during the 1990s. Some countries are making progress, but the stunning and sad fact is that the very poorest countries in the world, in general, are falling further and further behind in meeting those goals; and the rich countries that promised to help them to do more have really lost attention, I'm afraid, and are focusing so much on issues of terrorism, war and peace, and their own internal issues, that they're just not paying the attention that they promised to global poverty.

This is necessarily a contract between the rich and the poor?
When the goals were set at the Millennium Assembly and then followed up in several important gatherings in which the United States was the key participant – such as at the International Conference on Financing for Development in Monterrey, Mexico, in March 2002, which President Bush attended – the rich and the poor countries said, "We have to do this together." The rich countries acknowledged that the impoverished countries could not fight disease on their own or solve the problems of hunger on their own. They would need help – a lot more help than they receive today. This

commitment was put in a very specific promise: The rich countries would make concrete efforts toward raising their development assistance to 0.7% of their gross national product (GNP). But despite this promise, the situation is worsening throughout Africa and in many other impoverished regions of the world. I think it is really a terrible mistake on the part of the rich world not to be paying more attention to this. It hurts us in the end by contributing to global instability.

When you look at the numbers, however, it appears as though we have a real shot at ending poverty sometime soon.
Well, the crazy thing about all of this is that you'd think we'd be paying more attention to these life-and-death issues. There are millions of children dying every year of readily preventable or treatable conditions, like the nearly one million children dying of measles, even though there's a vaccine to stop it. There are nearly three million children dying of malaria, even though we have medicines that cure it. So you have this stunning challenge, but at the same time there are very specific, relatively straightforward interventions in a lot of cases that could address these problems. Poverty reduction is not rocket science, but the gap between where we are and what we could do if we fulfilled our promises is stunning.

With the work you've done at the Earth Institute at Columbia University, how have you come to know science and its importance in the role of ending global poverty?
It's a wonderful thing to talk to the real practitioners who can help guide you through these problems because sometimes we look at these issues and assume they're too big to confront. Then we talk to people who work on these problems for a living – the scientists, the technology experts, and the engineers – all of whom have very well-targeted, well-designed, often quite straightforward approaches to the problem.

Take the case of hunger, for example. I've been visiting farms in Africa in recent years with my colleague Dr. Pedro Sanchez, who is one of the world's leading soil scientists and winner of the 2002 World Food Prize. He is the co-coordinator of the Task Force on Hunger for the Millennium Project. When we go through the African farms in very poor areas of western Kenya or Ethiopia, for example, what look like incidental weeds or bushes to me are for Dr. Sanchez the solution to the problem. When the landscape is explained properly, Dr. Sanchez has made me understand that the farms in Africa are now operating on soils virtually depleted of nutrients. It's a cruelty that the women of Africa go out to perform backbreaking labor, farming their land, when the land doesn't even have nutrients anymore to sustain crops. Dr. Sanchez shows how various techniques can triple the yields of these poor farmers pretty quickly, in two or three years. These include straightforward chemical fertilizers but also approaches that he's been championing, like agroforestry (planting certain kinds of plants near the crops in order to replenish soil nutrients). The right kind of agricultural extension and some access to these technologies would result in huge increases of crop yields. Farmers would be able to feed themselves and their families. They would be healthier. They would be more resistant to disease. They would actually have a surplus to take to the market and engage in the economy.

The $25 billion that one would need to launch a serious attack on the killer diseases in the poorest countries is about one-thousandth of our annual income, or around ten cents for every hundred dollars of our income. In other words, utterly affordable.

How much would it take on the part of the rich world to turn the conditions around in the world's poorest areas?
When I chaired the Commission on Macroeconomics and Health for the World Health Organization (WHO) during 2000 and 2001, we did a very detailed analysis of what it would cost to address the key killers: AIDS, TB, malaria, vaccine-preventable diseases, respiratory infection in children, diarrheal disease that kills millions of children every year, unsafe childbirth, micronutrient deficiencies, and so on. A team at the London School of Hygiene and Tropical Medicine, led by professor Anne Mills, made a very thorough and detailed costing. What they found was that if you added all the costs, subtracted what poor countries spend on health today, and subtracted a plausible increase in health spending by those countries, then the remaining costs (which would have to be paid by the rich countries) would be on the order of about $25 billion per year. That actually used to sound like a big number, until now: We spent $87 billion in Iraq and Afghanistan and we gave away $250

billion in the Bush Administration's tax cuts. Twenty-five billion dollars a year, from the whole rich world, actually is not all that much, since the combined income of the entire rich world is about $25 trillion. The $25 billion that one would need to launch a serious attack on the killer diseases in the poorest countries is about one-thousandth of our annual income, or around ten cents for every hundred dollars of our income. In other words, utterly affordable. This would not break the bank. In fact, it would hardly be noticed in terms of our own income. But it would make a world of difference for the poorest countries in the world.

It's astounding to me that this message isn't more widely distributed and broadcast on a daily basis, as a reminder.
I've wondered why President Bush doesn't get up every day and state these things. Then I thought, "Why doesn't he say it even once?" The fact of the matter is that this message is not understood. The American people, who are very generous people, believe that we're already doing everything we can do. This is a big mistake. Their own leaders say, "Yes, we are generous people." But they don't tell them that because we're generous people we should be doing what we promised, not what we're pretending to do. The United States actually gives the smallest amount of development aid as a share of national income of any donor country in the world. We're just giving about 0.13% of our national income to development assistance. Most Americans think we give 5% of our income, or more! We only give 13 cents out of every hundred dollars, no more than that. That's not enough, nor is it anywhere close to what we've promised to do.

The result is tragedy in Africa, instability, and crises. These crises come back to haunt us in the form of international transmission of disease, bases for terrorism, U.S. involvement in emergency interventions like emergency food aid (which is quite expensive compared to what preventing the emergency would have cost), emergency military intervention, and now the rising cost of security spending in Africa. None of it makes sense if we would just follow through on what we promised, and took serious steps toward solving these problems.

Jeffrey Sachs is the director of the Earth Institute at Columbia University in New York.

Select bibliography

BOOKS

Ball, Philip. *The Ingredients: A Guided Tour of the Elements*, Oxford University Press, New York, 2002.
_____. *Made to Measure: New Materials for the 21st Century*, Princeton University Press, Princeton, NJ, 1999.

Barlow, Maude. *Blue Gold: The Global Water Crisis and the Commodification of the World's Water Supply*, rev. ed., International Forum on Globalization, San Francisco, Spring 2001.

Brand, Stewart. *The Clock of the Long Now: Time and Responsibility*, Basic Books, New York, 2000.

Burtynsky, Edward. *Manufactured Landscapes: The Photographs of Edward Burtynsky*, Yale University Press, New Haven, 2003.

De Soto, Hernando. *The Mystery of Capital: Why Capitalism Triumphs in the West and Fails Everywhere Else*, Basic Books, New York, 2000.

Der Derian, James. *Virtuous War: Mapping the Military-Industrial-media-entertainment Network*, Westview Press, Boulder, 2001.

Dyer, Gwynne. *Ignorant Armies: Sliding Into War in Iraq*, McClelland & Stewart, Toronto, 2003.
_____. *Future Tense: The Coming of World Order*, McClelland & Stewart, Toronto, 2004.

Dyson, Freeman. *Disturbing the Universe*, Harper Collins, New York, 1981.
_____. *Imagined Worlds (Jerusalem-Harvard Lectures)*, Harvard University Press, Cambridge, MA, 1998.
_____. *The Sun, The Genome, and The Internet: Tool of Scientific Revolutions*, Oxford University Press, Oxford, 1999.

Frankel, Felice. *Envisioning Science: The Design and Craft of the Science Image*, The MIT Press, Boston, 2002.

Frankel, Felice and George Whitesides. *On the Surface of Things*, Chronicle Books, San Francisco, 1997.

Henderson, Hazel. *Beyond Globalization: Shaping a Sustainable Global Economy*, Kumarian Press, Inc., Bloomfield, CT, 1999.

Homer-Dixon, Thomas. *The Ingenuity Gap: How Can We Solve the Problems of the Future?* Jonathan Cape, London, 2000.

Kroker, Arthur. *Technology and the Canadian Mind: Innis/McLuhan/Grant (New World Perspectives)*, Palgrave-Macmillan, Hampshire, UK, 1985.

Lessig, Lawrence. *Code: And Other Laws of Cyberspace*, Basic Books, New York, 2000.
_____. *Free Culture: How Big Media Uses Technology and the Law to Lock Down Culture and Control Creativity*, Penguin Books, New York, 2004.

McDonough, William and Michael Braungart. *Cradle to Cradle: Remaking the Way We Make Things*, North Point Press, New York, 2002.

Melman, Seymour. *The Permanent War Economy*, Simon & Schuster, Glasgow, 1985.

Papanek, Victor. *Design for the Real World: Human Ecology and Social Change*, Thames and Hudson, London, 1985.

Ridley, Matt. *Genome: The Autobiography of a Species in 23 Chapters*, Fourth Estate, London, 2000.

Rifkin, Jeremy. *The Hydrogen Economy: The Next Great Economic Revolution*, Jeremy P. Tarcher, New York, 2002.

Smith, Dan. *The Penguin State of the World Atlas*, 7th ed., Penguin Books, New York, 2003.

Smith, Dan. *The Penguin Atlas of War and Peace*, Penguin Books, New York, 2003.

Sterling, Bruce. *Tomorrow Now: Envisioning the Next 50 Years*, Random House, New York, 2003.

Sutcliffe, Bob. *100 Ways of Seeing an Unequal World*, Zed Books, London, 2002.

Thacker, Eugene. *Biomedia*, University of Minnesota Press, Minneapolis, MN, 2004.

Toffler, Alvin. *The Third Wave*, Bantam Books, New York, 1981.

Winchester, Simon. *The Meaning of Everything: The Story of the Oxford English Dictionary*, Oxford University Press, Oxford, 2003.

PERIODICALS

Arnst, Catherine, with John Carey, "Biotech Bodies," *BusinessWeek*, July 27, 1998, pp. 56–63.

Becker, Jasper. "World's third largest river starts to rise by 400ft to create the Great Wall of Water," *The Independent*, UK, June 1, 2003.

Burke, Monte. "Plastic Man," *Forbes*, December 23, 2002, pp. 296–98.

Chase, Marilyn. "African girls taught to say no to 'sugar daddies,'" *The Wall Street Journal*, February 25, 2004.

Cook, Lynn J. "Millions Served," *Forbes*, December 23, 2002, pp. 302–4.

Foster, Ian, Carl Kesselman, and Steven Tuecke. "The Anatomy of the Grid: Enabling Scalable Virtual Organizations" in *Intl J. Supercomputer Applications*, 2001, pp. 1–25.

Gust, Devens, Thomas A. Moore, and Ana L. Moore, "Mimicking Photosynthetic Solar Energy Transduction" in *Acc. Chem. Res.* 34 (2001): 40–48.

Wong, Kate. "Gecko-Inspired Adhesive Sticks it to Traditional Tape," *Scientific American*, June 4, 2003, online edition.

Langer, Robert S., and Joseph P. Vacanti, "Tissue Engineering: The Challenges Ahead," *Scientific American*, April 1999, pp. 86–89.

Langer, Robert, and Nicholas A. Peppas. "Advances in Biomaterials, Drug Delivery, and Bionanotechnology," *AIChE Journal* 49:12 (December 2003): 2990–3006.

Mooney, David J., and Antonios G. Mikos, "Growing New Organs," *Scientific American*, April 17, 1999, pp. 59–65.

Rybczynski, Witold. "Living Smaller," *The Atlantic*, February 1991, pp. 64–78.

Sachs, Jeffrey. "A Simple Plan to Save the World," *Esquire*, May 2004, pp. 125–29, 146–47.

Sachs, Jeffrey. "The G8 Must Fund the War Against Poverty," *Financial Times*, June 7, 2004, p. 17.

Sachs, Jeffrey. "Doing the sums in Africa," *The Economist*, May 20, 2004.

Thacker, Eugene. "What is Biomedia?" *Configurations* 11:1 (Winter 2003): 47–49.

WEBSITES

Addams, Cheryl. "A World Service for the Boeing 737," Boeing, www.boeing.com/commercial/news/feature/737.html.

Baum, Rudy. "Nanotechnology," *Chemical & Engineering News*, December 1, 2003, pubs.acs.org/cen/coverstory/8148/8148counterpoint.html.BP, "BP Hornchurch Connect is a holistic approach to sustainable retailing," BP Hornchurch Connect, subsites.bp.com/centres/press/hornchurch/highlights/ factsheet.asp.

Chen, Daniel. "Area of Earth's Land Surface," (edited by Glenn Elert), The Physics Factbook, hypertextbook.com/facts/2001/DanielChen.shtml.

ChinaOnline. "Three Gorges Dam Project," www.chinaonline.com/refer/ministry_profiles/threegorgesdam.asp.

Democratic Socialists of America. "The Wal-Mart Revolution," www.dsausa.org/lowwage/walmart/why_walmart.html.

Demographia. "Large International Urbanized Areas; Population, Land Area & Density," www.demographia.com/db-intl-ua2001.htm.

Discoe, Ben. "Dymaxion car chronology," Washed Ashore, www.washedashore.com/projects/dymax/chronology.html.

Foresight Institute. "Nanotechnology Pioneer Calms Fears of Runaway Replicators," Press Center, www.foresight.org/.

Gilbert, Alorie. "Retail's Super Supply Chains," *InformationWeek*, www.informationweek.com/808/walmart.htm.

Guinness World Records. "Least Dense Solids," Science and Technology, Amazing Science, www.guinnessworldrecords.com/.

Instituto de Pesquisa e Planejamento Urbano de Curitiba. "Thinking the city," IPPUC, www.ippuc.pr.gov.br.

Institute for Scientific Research, Inc. "The Space Elevator," www.isr.us/SEHome.asp.

Jet Propulsion Laboratory, California Institute of Technology. "Ideas that Gel," JPL, February 11, 2002, www.jpl.nasa.gov/news/features.cfm?feature=490.

Luo, Annie. "Three Gorges: 'A World So Changed,'" www.worldpress.org/Asia/1245.cfm, July 8, 2003.

Lyon, Barret. "Project History," The Opte Project, www.opte.org/.

May, Paul. "What is Diamond CVD Technology?" Bristol University CVD Diamond Group, www.chm.bris.ac.uk/~paulmay/diamhome.htm.

Melman, Seymour. "How George Bush, His Congress and Pentagon are Looting Our Cities, Robbing our People & Stealing From Our Children," *War Economy Papers #4*, April 17, 2003, www.aftercapitalism.com.

Negative Population Growth. "Sprawl & Population Growth," NPG Fact Sheet, www.npg.org/factsheets/sprawl.html.

Neighborhood Electric Vehicle Company. "Gizmo affordable eco-fun," NEVCO, www.nevco.com/.

Railway Technical Research Institute. "Overview of Maglev R&D," www.rtri.or.jp/rd/maglev/html/english/maglev_frame_E.html.

Roy, Arundhati. "Do turkeys enjoy thanksgiving?" *Freidenspolitischer Ratschlag*, Mumbai, www.uni-kassel.de/fb10/frieden/themen/Globalisierung/roy2-orig.html.

Rule, Audrey C. "The Rhyming Peg Mnemonic Device Applied to Learning the Mohs Scale of Hardness," Department of Curriculum and Instruction, State University of New York at Oswego, www.nagt.org/Nov03/RULE_05_03.pdf.

Ryzhov, Dr. Valentin. "Method for the production of single-crystal c-BN powders," Institute for High Pressure Physics, Russian Academy of Sciences, www.hppi.troitsk.ru/aboutus.htm.

Shell Renewables. "A powerful force for tomorrow's generation," Shell WindEnergy, shell.com/renewables.

Simulation, Modeling & Control. "SiMiCon Rotor Craft – SRC," SiMiCon, www.simicon.no/.

Team 24355 and Kayotic Development. "Conceiving a Clone," library.thinkquest.org/24355/.

The Earth Simulator Center. "Birth of Earth Simulator," Earth Simulator, www.es.jamstec.go.jp/esc/eng/.

U.S. Department of Energy, Office of Energy Efficiency and Renewable Energy. "Wind Farms and Wind Farmers," Consumer Energy Information: EREC Reference Briefs, www.eere.energy.gov/consumerinfo/refbriefs/ad2.html.

OTHER SOURCES

Ashoka: Innovators for the Public. "The Citizen Sector Transformed," PowerPoint presentation, Said University, Oxford University, March 30, 2004.

Ball, Philip. "Materials of the Future," a chapter for the UNESCO Encyclopedia of Life Support Systems, 2001.

Biberdorf, Curt. "Future Warrior," *The Warrior*, U.S. Army Soldier Systems Center newsletter, July–August, 2003.

Biddle, Stephen. "Afghanistan and the Future of Warfare: Implications for Army and Defense Policy," Army War College Strategic Studies Institute, Carlisle Barracks, PA, November 2002.

Broughton, John. "The Bomb's-Eye View: Smart Weapons and Military T.V.," manuscript.

The Chaordic Commons. "What is VISA?" PowerPoint presentation, 2003.

The Natural Step. "Bank of America Case Study," © 2003.

The Natural Step. "The Home Depot Case Study," © 2003.

The Natural Step. "McDonald's Corporation Case Study," © 2003.

The Natural Step. "Nike, Inc. Case Study," © 2003.

The Natural Step. "Starbucks Coffee Company Case Study," © 2003.

The Natural Step. "Whistler, British Columbia Case Study," © 2003.

The Natural Step. "IKEA: The Natural Step Organizational Case Summary," © 2003.

Pearson, Lester Bowles. Nobel Prize Lecture, University of Oslo, December 11, 1957.

Sachs, Jeffrey D. "Ending Global Poverty" (transcript), Hilton Foundation Conference, New York, September 24, 2003.

_____. "Achieving the Millennium Development Goals: Health in the Developing World," speech at the Second Global Consultation of the Commission on Macroeconomics and Health, Geneva, October 29, 2003.

SLAC–Fermilab. "Grid2003: A Persistent U.S. Science Grid," PowerPoint presentation, courtesy Argonne National Laboratory, The University of Chicago.

Smalley, R. E. "Our Energy Challenge," PowerPoint presentation, Columbia University, September 23, 2003.

World Business Council for Sustainable Development. "Mobility 2001: World Mobility at the End of the Twentieth Century and its Sustainability," member report, 2001.

World Design Science Decade 1965–1975. Phase I (1963), Inventory of World Resources Human Trends and Needs, World Resources Inventory, Southern Illinois University, Carbondale, IL.

World Design Science Decade 1965–1975. Phase I (1965), Document 4, "The Ten Year Program," World Resources Inventory, Southern Illinois University, Carbondale, IL.

World Game Studies Workshop. "Explorations Into Alternatives for Spaceship Earth," June 27 to July 16, 1971, Department of Design, Southern Illinois University, Carbondale, IL.

World Health Organization. "World report on road traffic injury prevention," April 2004.

INTERVIEWS

BY LORRAINE GAUTHIER

James H. Korris, Marina del Rey, CA, November 2003.

BY JENNIFER LEONARD

Philip Ball, telephone interview live-to-air, CIUT 89.5 FM (University of Toronto), March 9, 2004.

Janine Benyus, telephone interview live-to-air, CIUT 89.5 FM (University of Toronto), October 14, 2003.

Stewart Brand, telephone interview (pre-recorded), CIUT 89.5 FM (University of Toronto), February 2, 2004.

John Broughton, telephone interview (pre-recorded), CIUT 89.5 FM (University of Toronto), June 3, 2004.

Stephen Browne, telephone interview live-to-air, CIUT 89.5 FM (University of Toronto), June 15, 2004.

Carol Burns, telephone interview live-to-air, CIUT 89.5 FM (University of Toronto), February 24, 2004.

Bill Buxton, in-person interviews, November 18, 2003; February 6, 2004.

James Der Derian, telephone interview live-to-air, CIUT 89.5 FM (University of Toronto), June 1, 2004.

Bill Drayton, telephone interview (pre-recorded), CIUT 89.5 FM (University of Toronto), June 3, 2004.

Gwynne Dyer, telephone interview (pre-recorded), CIUT 89.5 FM (University of Toronto), May 21, 2004.

Freeman Dyson, telephone interview live-to-air, CIUT 89.5 FM (University of Toronto), November 4, 2003.

Aleksandr Ermishin, Ekip Aviation Concern, email interview, March 23, 2004.

Kim Forte, EnviroMission, email interview, March 24, 2004.

Ian Foster, telephone interview live-to-air, CIUT 89.5 FM (University of Toronto), December 9, 2003.

Felice Frankel, telephone interview live-to-air, CIUT 89.5 FM (University of Toronto), February 10, 2004.

Ashok Gadgil, telephone interview live-to-air, CIUT 89.5 FM (University of Toronto), November 11, 2003.

Steve Goldstein, University of Michigan, email interview, June 15, 2004.

Catherine Gray, telephone interview live-to-air, CIUT 89.5 FM (University of Toronto), April 13, 2004.

Hazel Henderson, telephone interview live-to-air, CIUT 89.5 FM (University of Toronto), September 9, 2003.

Richard Jefferson, CAMBIA, email interview, June 23, 2004.

Dean Kamen, telephone interview live-to-air, CIUT 89.5 FM (University of Toronto), September 30, 2003.

Arthur Kroker, telephone interview (pre-recorded), CIUT 89.5 FM (University of Toronto), June 3, 2004.

Bob Langer, telephone interview live-to-air, CIUT 89.5 FM (University of Toronto), January 13, 2004.

Jaime Lerner, telephone interview (pre-recorded), CIUT 89.5 FM (University of Toronto), April 15, 2004.

Lawrence Lessig, telephone interview live-to-air, CIUT 89.5 FM (University of Toronto), January 20, 2004.

David Malin, telephone interview, May 3, 2004.

Susan McCouch, Cornell University, email interview, June 11, 2004.

Michael McDonough, telephone interview live-to-air, CIUT 89.5 FM (University of Toronto), November 18. 2003.

William McDonough, telephone interview (pre-recorded), CIUT 89.5 FM (University of Toronto), March 2, 2004.

Seymour Melman, telephone interview live-to-air, CIUT 89.5 FM (University of Toronto), December 16, 2003.

Michelle Murray, Enwave, telephone interview, February 23, 2004.

Tapio Niemi, Turbostove, email interview, March 28, 2004.

Hal Pierce, NASA, email interview, April 28, 2004.

Nancy Padian, telephone interview live-to-air, CIUT 89.5 FM (University of Toronto), September 23, 2003.

Matt Ridley, telephone interview live-to-air, CIUT 89.5 FM (University of Toronto), January 6, 2004.

Jeffrey Sachs, telephone interview live-to-air, CIUT 89.5 FM (University of Toronto), March 16, 2004.

John Santini, MicroCHIPS, Inc., email interview, February 26, 2004.

Neil Singer, Sandia, email interview, April 27, 2004.

Rick Smalley, telephone interview live-to-air, CIUT 89.5 FM (University of Toronto), November 25, 2003.

Hernando de Soto, telephone interview live-to-air, CIUT 89.5 FM (University of Toronto), October 21, 2003.

Richard Stallman, videoconference interview, July 29, 2003.

Bruce Sterling, telephone interview live-to-air, CIUT 89.5 FM (University of Toronto), December 2, 2003.

Dan Sturges, email interview, March 15, 2004.

Eugene Thacker, telephone interview live-to-air, CIUT 89.5 FM (University of Toronto), June 8. 2004.

John Todd, Ocean Arks International, email interview, June 13, 2004.

Scott White, University of Illinois at Urbana-Champaign, email interview, May 4, 2004.

GRAPH SOURCES

8.05 Melman, Seymour. "How George Bush, His Congress and Pentagon Are Looting Our Cities, Robbing our People & Stealing From Our Children," *War Economy Papers*, #4, April 17, 2003, www.aftercapitalism.com.

11.00 Sutcliffe, Bob. "Human Development Index," graph 122 in *100 Ways of Seeing an Unequal World*, Zed Books, Ltd., London, 2001.

11.01 to 11.04 "The Citizen Sector Transformed," Ashoka PowerPoint presentation, Said Business School, Oxford University, March 30, 2004.

11.05 to 11.08 UNDP, ICT.

11.09 to 11.12 Women's Global Health Imperative.

11.13 to 11.16 From the following articles by Jeffrey Sachs: "A Simple Plan to Save the World," *Esquire*, May 2004; "The G8 Must Fund the War Against Poverty," *Financial Times*, June 7, 2004; "Doing the sums in Africa," *The Economist*, May 20, 2004.

ILLUSTRATION CREDITS

COVER
Courtesy Barrett Lyon, The Opte
Project

ENDPAPERS–PAGE 11
© AFP/Robert Laberge: Endpaper
© Robert Polidori: Endpaper–p.1
Courtesy FEMA News Photo: 2-3
Courtesy Rick Dove, Waterkeeper
Alliance (www.neuseriver.com):
4-5
Courtesy FEMA News
Photo/Andrea Booher: 6-7
© AFP/Jon Levy: 8-9
Courtesy U.S. Geological
Survey/David Carver: 10-11

URB
1.00 Courtesy Eva Kato
1.01 © Wichita County Historical
Museum
1.02 Courtesy Structural Insultated
Panel Association (SIPA) &
Enercept, Inc.
1.03 Courtesy BoKlok
1.04 Courtesy National Archives
and Records Administration
1.05 iStock Photo/© Rich Cutter
1.06 Courtesy Keo Iv-Ty
1.07 Courtesy Stéphane Tondu-
Tsuya
1.08 Courtesy Augusto Tavares
Rosa Marcacini
1.09 Courtesy Martijn Leensen
1.10 iStock Photo/© Mak Takman
1.11 Courtesy Michael McDonough
Architect
1.12 Courtesy Jeffrey "Boof"
Bordman
1.13 Courtesy Jennifer Leonard
1.14 Courtesy NASA EOS, Terra
satellite

MOV
2.00 © Transrapid International
2.01 Sara Moss
2.02 © Segway LLC
2.03 Courtesy The Estate of R.
Buckminster Fuller
2.04 Courtesy Chris Owen, Planet
Sinclair
(www.nvg.ntnu.no/sinclair)
2.05 Courtesy Hypercar, Inc./Norm
Clasen
2.06 Courtesy Bombardier Recreational
Products
2.07 Courtesy Wheelsurf Sport
Ltda.
2.08 © Neighborhood Electric
Vehicle Company
2.09 Courtesy Dan Sturges
2.10 Courtesy Matthew Mishrikey
2.11 to 2.15 Courtesy IPPUC,
Curitiba, Brazil/Carlos Ruggi
2.16 © Transrapid International
2.17 Courtesy Timothy Sullivan
2.18 © Bruce Mau Design, Inc.
(design by Christopher Bahry)
2.19 © CargoLifter AG i.I.
2.20 © Institute for Scientific
Research, Inc.
2.21 © EKIP Aviation Concern
2.22 © EKIP Aviation Concern
2.23 © Avro Canada
(www.avroland.ca)
2.24 © SiMiCon

ENE
3.00 Courtesy Solar & Heliospheric
Observatory (SOHO)
3.01 © Guttierez+Portefaix
3.02 Courtesy U.S. Bureau of
Reclamation
3.03 © Enwave District Energy
Limited
3.04 Courtesy Jörg Schlaich,
Schlaich Bergermann und
Partner
3.05 Courtesy Ian Bryden,
The Robert Gordon University
3.06 Courtesy U.S. Department of
Energy/Sandia National
Laboratory

3.07 © BP p.l.c.
3.08 © Stuart Energy Systems
Corporation
3.09 © Shell Hydrogen
3.10 and 3.11 Courtesy Tapio
Niemi/Turbostove
3.12 © Guttierez+Portefaix
3.13 Courtesy U.S. Department of
Energy
3.14 © Patrice Hanicotte
3.15 to 3.17 Courtesy Solar
Electric Light Fund
3.18 Courtesy Ana Moore,
University of Arizona
3.19 Courtesy Solar Electric Light
Fund
3.20 to 3.22 Courtesy The Smalley
Group, Rice University

INF
4.00 and 4.01 Courtesy Barrett
Lyon, The Opte Project
4.02 © Larry Ewing
4.03 Courtesy Free Software
Foundation
4.04 Courtesy Secretary to the
Delegates, Oxford University
Press
4.05 © Napster
4.06 Courtesy CERN
4.07 Courtesy The National
Center for Supercomputing
Applications
4.08 Courtesy Bruce Mau Design, Inc.
4.09 Courtesy Fritz Hasler and Hal
Pierce/NASA Goddard Space
Flight Center
4.10 to 4.13 Courtesy Ozone
Processing Team/NASA
Goddard Space Flight Center
4.14 Courtesy NASA Earth
Observatory Satellite/
© JAMSTEC/Earth Simulator
Center
4.15 © JAMSTEC/Earth Simulator
Center

IMA
5.00a Courtesy Fermi National
Accelerator Laboratory
5.00b Courtesy NASA, ESA,
S. Beckwith (STScI) and the
HUDF Team
5.00 © Anglo-Australian
Observataory, photograph by
David Malin
5.01 Courtesy ESA, NASA,
J.P.Knieb (Caltech/Observatoire
Midi-Pyrénées) and R. Ellis
(Caltech)
5.02 Courtesy NOAA/Peter Dodge,
AOML Hurricane Research
Division
5.03 Courtesy NRAO/AUI
5.04 © 2000 General Electric,
Courtesy U.S. National
Library of Medicine; Courtesy
Volume Graphics GmbH
Heidelberg, Germany
(www.volumegraphics.com)
5.05 © Project VOXEL-MAN,
Institute of Medical Informatics,
University Hospital Hamburg-
Eppendorf, Germany
5.06 © Steve Grewell
5.07 © Kenneth S. Zirkel
5.08 Courtesy Trinity Route Survey
Office, Maritime Forces
Atlantic, Halifax
5.09 Courtesy NASA/JPL-Caltech;
NASA Landsat Project Science
Office and USGS EROS Data
Center
5.10 © Sierra Pacific Corp.
(www.nationalinfrared.com);
© Denise McQuillen;
NASA/JPL/Cornell/ASU;
Michael Hauser (STScI), the
COBE/DIRBE Science Team,
and NASA
5.11 See Acknowledgments for the
long list of contributors. Special
thanks go to NISEE (UC
Berkeley), NOAA's Office of
Response and Restoration, and
FEMA (row 1); Surveillance
Camera Players (row 6); Jamie

Osborne (row 8); and Caleb
Waldorf ("Operation: Iraqi
Freedom II" art project, rows
17 and 18).
5.12 Courtesy SOHO (ESA &
NASA); NASA/JPL-Caltech;
NASA/JPL/Space Science
Institute; ESA/NASA/University
of Arizona SST/Royal Swedish
Academy of Sciences
5.13 © American Science and
Engineering, Inc.; Courtesy
Philips Research, Hamburg
Germany; © Matthew Canada,
© Matthew Buchanan, © Erin
Walsh, Koert van Kleef,
© Bernd Klumpp, Virginia
Hilyard; Courtesy Joe Kniss;
Courtesy Volume Graphics
GmbH Heidelberg, Germany;
Courtesy Dr. Steve Parker of the
Scientific Computing and
Imaging (SCI) Institute at the
University of Utah
5.14 Courtesy of DigiMorph.Org
and The University of Texas at
Austin. Based on scans from the
High-Resolution X-ray
Computed Tomography Facility
at the University of Texas at
Austin
5.15 © MicroAngela; Chad Mirkin;
© Daniel Mathys; © Dee Breger
5.16 Courtesy CERN Geneva
5.17 Courtesy Lawrence Berkeley
National Laboratory
5.18 © International College of
Nuclear Medicine Physicians
5.19 Image reproduced by
permission of IBM Research.
Unauthorized use not permitted.
5.20 Courtesy K. Sahu, M. Livio, L.
Petro, D. Macchetto, STScI and
NASA
5.21 © Felice Frankel

MAR
6.00 © Guttierez+Portefaix
6.01 © Wal-Mart Stores, Inc.
6.02 © FedEx Corp.
6.03 Bruce Mau Design, Inc.
(design by Christopher Bahry)
6.04 © Texas Instruments, Inc.
6.05 and 6.06 Courtesy·Russell W.
Belk, University of Utah
6.07 Bruce Mau Design, Inc.
(design by Christopher Bahry)
6.08 Bruce Mau Design, Inc.

MAT
7.00 Courtesy Dr. Robert S.
Langer, Massachusetts Institute
of Technology and Dr. Joseph P.
Vacanti, Massachusetts General
Hospital
7.01 Courtesy Institute for High
Pressure Physics, Russian
Academy of Sciences
7.02 Courtesy Mehmet Sarikaya
7.03 and 7.04 Courtesy Dr. Paul
May, School of Chemistry,
University of Bristol, UK
7.05 Courtesy NSF/Dr. Laurence
Garvie
7.06 Courtesy Fritz Vollrath
7.07 and 7.08 © Nexia
Biotechnologies, Inc./Sean
O'Neill
7.09 iStock/© Kevin Tate
7.10 and 7.11 Courtesy Andre
Geim
7.12 to 7.14 Courtesy
Jet Propulsion Laboratory
7.15 © Roland McMahon
7.16 © IBM Almeda
7.17 © The Foresight Institute
7.18 Courtesy Sandia National
Laboratories, SUMMiT™
Technologies
7.19 Courtesy Sir Harold Kroto,
University of Sussex
7.20 and 7.21 © Felice Frankel
7.22 Courtesy Dr. Eric N. Brown,
Autonomic Materials Group,
Beckman Institute for Advanced
Science and Technology,
University of Illinois at Urbana-

Champaign
7.23 Courtesy George Lisensky,
Chemistry Department, Beloit
College
7.24 © mnemoScience GmbH

MIL
8.00 Courtesy U.S. Air Force;
photo by Staff Sgt. Chenzira
Mallory
8.01 © 2004 Eames Office
(www.eamesoffice.com)
8.02 © Sarah Underhill,
U.S. Army Soldier Systems
Center, Natick, Massachusetts
8.03 Institute for Creative
Technologies, University of
Southern California
8.04 Clockwise from top left:
Common Remotely Operated
Weapon Station (CROWS)
courtesy Program Executive
Office (PEO) Soldier website,
U.S. Army; Hi-Viz™
International Orange courtesy
Camelbak® Maximum Gear;
Gore-Tex courtesy U.S.
Department of Defense Logistics
Agency (DLA) website; Skydex
helmet courtesy Schutt Sports
DNA™; Supersoaker © 2003
Hasbro, Inc.; Jolt Caffeine-
Energy gum © 2003
GumRunners LLC; Bruce Mau
Design, Inc.; © Sarah Underhill,
U.S. Army Soldier Systems
Center, Natick, Massachusetts

MAN
9.00 all three opening images,
© Edward Burtynsky
9.01 © Susan Dobson
9.02 © Maconochie Photography
9.03 Courtesy Ocean Arks
International
9.04 © Young & Rubicam
9.05 © Herman Miller, Inc.
9.06 to 9.08 © 3D Systems
9.09 and 9.10 © IBM

LIV
10.00 Courtesy Lawrence Berkeley
National Lab
10.01 Courtesy Dr. Fu Xiqin, Dr.
Sri Koerniati, Dr. Andrzej Kilian
and Dr. Richard Jefferson, all of
CAMBIA
10.02 and 10.03 Courtesy Susan
McCouch, Department of Plant
Breeding, Cornell University
10.04 and 10.05 Courtesy Florence
Wambugu, Africa Harvest
(www.ahbfi.org)
10.06 Courtesy Peter Beyer,
Center for Applied Biosciences,
University of Freiburg, Germany
10.07 Courtesy Roslin Institute
10.08 Adi Nes, Untitled (The
Featherless Chicken), 2002,
Color photograph 60 x 90 cm;
Courtesy The New York Times
Magazine and Dvir Gallery,
Tel Aviv
10.09 Courtesy Dr. Robert S.
Langer, Massachusetts Institute
of Technology and Dr. Joseph P.
Vacanti, Massachusetts General
Hospital
10.10 © Yadong Wang
10.11 © Organogenesis Inc.
10.12 © MicroCHIPS, Inc.
10.13 and 10.14 Courtesy Ashok
Gadgil, Lawrence Berkeley
National Lab
10.15 © Bruce Mau Design, Inc.
10.16 and 10.17 Courtesy Ocean
Arks International

WP
All graphs: See bibliography,
previous page.

FINAL SPREAD
Courtesy Alex Adai, University of
California San Francisco

Index

To learn more about Phaidon, to keep up to date with our publications, to sign up to receive our newsletter, and to benefit from special promotions, visit us at

www.phaidon.com

All of these changes are leading us to an inevitable conclusion, a place we must go, if we can – a place that promises peace and prosperity within a sustainable future.

If we are to survive as a life form on the planet, at more or less the scale of our present occupation, there is no other way. Collectively, we must come to the realization that there is no exterior to our ecology. There is only one environment and everything is entered on the balance sheet. Every positive. Every negative. Everything counts. There can only be true prosperity if it is global prosperity, and we can only count our wealth in peace when we count it together.

The good news is that there is evidence all around us that not only are we as a global culture committed